C000291979

BRITISH BUTTERFLIES

[illegible]

TRANSFORMATIONS.

British
Butterflies
and their
Transformations.
LONDON

BRITISH BUTTERFLIES

AND THEIR

TRANSFORMATIONS.

ARRANGED AND ILLUSTRATED IN A SERIES OF PLATES BY

H. N. HUMPHREYS, ESQ.

WITH CHARACTERS AND DESCRIPTIONS BY

J. O. WESTWOOD, ESQ., F.L.S.,

SEC. OF THE ENTOMOLOGICAL SOCIETY

LONDON
WILLIAM SMITH, 113, FLEET STREET
MDCCCXLI

SMITHSONIAN
MAY 26 1955
LIBRARY

CONTENTS

b

LIST OF PLATES

No. I.] [PRICE 2s. 6d.

BRITISH INSECTS AND THEIR TRANSFORMATIONS.

BRITISH BUTTERFLIES

AND

Their Transformations,

ARRANGED AND ILLUSTRATED IN A SERIES OF PLATES

BY H. N. HUMPHREYS, ESQ.

WITH CHARACTERS AND DESCRIPTIONS

BY J. O. WESTWOOD, ESQ., F.L.S.,

SEC. OF THE ENTOMOLOGICAL SOCIETY, ETC. ETC.

LONDON:
WILLIAM SMITH, 113, FLEET STREET.

MDCCCXL.

WORKS PUBLISHED BY WILLIAM SMITH, 113, FLEET STREET.

SMITH'S STANDARD LIBRARY.

In medium 8vo, uniform with Byron's Works, &c

Works already Published

THE LADY OF THE LAKE 1s
THE LAY OF THE LAST MINSTREL. 1s
MARMION. 1s 2d
THE VICAR OF WAKEFIELD. 1s
THE BOROUGH BY THE REV GEORGE CRABBE 1s 4d
THE MUTINY OF THE BOUNTY 1s 4d.
THE POETICAL WORKS OF H KIRKE WHITE 1s
THE POETICAL WORKS OF ROBERT BURNS 2s 6d
PAUL AND VIRGINIA, THE INDIAN COTTAGE, AND ELIZABETH. 1s 6d
MEMOIRS OF THE LIFE OF COLONEL HUTCHINSON (Governor of Nottingham Castle during the Civil War) By his Widow Mrs Lucy Hutchinson. 2s 6d
THOMSON'S SEASONS, AND CASTLE OF INDOLENCE 1s
LOCKE ON THE REASONABLENESS OF CHRISTIANITY 1s
GOLDSMITH'S POEMS AND PLAYS. 1s 6d
KNICKERBOCKER'S HISTORY OF NEW YORK 2s 3d
NATURE AND ART By Mrs Inchbald 10d
SCHILLER'S TRAGEDIES THE PICCOLOMINI, and THE DEATH OF WALLENSTEIN 1s 8d
ANSON'S VOYAGE ROUND THE WORLD 2s 6d
THE POETICAL WORKS OF GRAY AND COLLINS 10d
THE LIFE OF BENVENUTO CELLINI 3s
HOME By Miss Sedgwick 9d
THE VISION OF DON RODERICK, AND BALLADS By Sir Walter Scott 1s
TRISTRAM SHANDY By Laurence Sterne 3s
STEPHENS'S INCIDENTS OF TRAVEL IN EGYPT, ARABIA PETRÆA, AND THE HOLY LAND 2s 6d
STEPHENS'S INCIDENTS OF TRAVEL IN GREECE, TURKEY, RUSSIA, AND POLAND 2s 6d
BEATTIE'S POEMS AND BLAIR'S GRAVE 1s
THE LIFE AND ADVENTURES OF PETER WILKINS 2s 6d.
POPE'S HOMER'S ILIAD 3s
UNDINE A Miniature Romance, from the German 9d.
ROBIN HOOD, reprinted from Ritson 2s 6d
LIVES OF DONNE, WOTTON, HOOKER, HERBERT, AND SANDERSON By Izaak Walton 2s 6d
LIFE OF PETRARCH By Mrs Dobson 3s
GOLDSMITH'S CITIZEN OF THE WORLD 2s. 6d
MILTON'S PARADISE LOST 1s 10d
MILTON'S PARADISE REGAINED AND MINOR POEMS 2s
RASSELAS. By Dr Johnson. 9d.
ALISON'S ESSAYS ON TASTE 2s 6d

THREE VOLUMES ARE NOW ARRANGED ON THE FOLLOWING PLAN —

ONE VOLUME OF "POETRY,"

CONTAINING

SCOTT'S LAY OF THE LAST MINSTREL
SCOTT'S LADY OF THE LAKE
SCOTT'S MARMION
CRABBE'S BOROUGH
THOMSON'S POETICAL WORKS
KIRKE WHITE'S POETICAL WORKS
BURNS'S POETICAL WORKS

ONE VOLUME OF "FICTION,"

CONTAINING

NATURE AND ART By Mrs INCHBALD
HOME By Miss SEDGWICK
KNICKERBOCKER'S HISTORY OF NEW YORK By WASHINGTON IRVING
PAUL AND VIRGINIA By ST PIERRE
THE INDIAN COTTAGE By ST PIERRE
ELIZABETH By MADAME COTTIN
THE VICAR OF WAKEFIELD By OLIVER GOLDSMITH
TRISTRAM SHANDY By LAURENCE STERNE

ONE VOLUME OF "VOYAGES AND TRAVELS,"

CONTAINING

ANSON'S VOYAGE ROUND THE WORLD
BLIGH'S MUTINY OF THE BOUNTY
STEPHENS'S TRAVELS IN EGYPT, ARABIA PETRÆA, AND THE HOLY LAND
STEPHENS'S TRAVELS IN GREECE, TURKEY, RUSSIA, AND POLAND

These Volumes may be had separately, very neatly bound in cloth, with the contents lettered on the back,

PREFACE.

During a recent tour through Italy I first conceived a predilection for the study of entomology. Early in the Italian spring, in the months of March and April, after a winter's residence in Rome, my favourite rambles were over the desert yet beautiful Campagna; and in these walks my attention was actively aroused by the profusion and variety of insect life, particularly of glittering butterflies, that in those early months already flitted over the flowery waste. As I stepped among tufts of the Alpine anemony, the crimson cyclamen, or purple squill, crowds of painted insects arose at every tread, as though a passing gust of wind had suddenly scattered a cloud of the many-coloured petals of the crushed flowers to the breeze. Later in the season the numbers still increased, and their brilliancy and novelty soon determined me to attempt the formation of a collection, reserving the classification and study till my return home; when I discovered that many beautiful species of Lepidoptera which I had deemed novelties were well known as indigenous to our own island; where, however, their comparatively unfrequent appearance had not forced them into notice, whilst in Italy their profusion had compelled attention. Such was the case with Papilio Machaon (the swallow-tail butterfly) as common on the Roman Campagna as the cabbage-white in our gardens. Mancipium Daplidice (the Bath white) and Pieris Cratægi (the black-veined white) were still more numerous; whilst the whole Campagna about mid-day received quite a golden hue from the rich orange colours of Gonapteryx Cleopatra and Colias Edusa, both of which were in such profusion that I actually took above twenty specimens of the latter species at once, upon a gigantic thistle on the road to Tivoli.

Upon my return to England I found the arrangement and description of my collection not so easy a matter as I had anticipated, for want of some popular yet comprehensive and complete work upon the subject; and I was only eventually enabled to make myself acquainted with even the British species in their different stages by reference to expensive foreign works. Feeling thus the want of some popular work in which the transformations of British Lepidoptera were accurately described and developed, with accurate portraits of each insect in its three great stages—* the caterpillar, the chrysalis, and the butterfly or moth, I planned the present work with the view of supplying the deficiency I had so much felt myself; not confining myself to Lepidoptera but extending the plan to all British insects, many of whose transformations are perhaps even more wonderful than those of

* Several works exist which are nearly complete with reference to the imago or complete insect, but none representing all the British species in the three stages of larva, pupa, and imago; the figures in Stephens, Curtis, or Donovan, being only selections.

butterflies, and with the able assistance of Mr Westwood, who has kindly undertaken to describe the characters &c &c, I hope to make the present a more complete work than any that has hitherto appeared in this country upon the subject, whilst its price will make it attainable by the great mass of the public

In this place it is usual to put forth some argument in favour of the study of such subjects as the book treats of, but it seems scarcely necessary to urge anything in favour of the delightful study of entomology The great beauty of many tribes of insects, their wonderful and minute organisation, their extraordinary metamorphoses, and the links they add to the chain of created beings, appear to form an all-sufficient attraction

All must acknowledge the desolateness and vacancy of the mind, which, placed among the treasures of a splendid library, is unable to taste the rich fruits of reason that are piled around—unable to read Yet, without being acknowledged or felt by the great mass of society, just such is the situation of one who has never awakened to the wonders of natural history, when he finds himself in the woods or fields He cannot read in the beautiful book of nature, when in the summer it opens its brightest leaves, illuminated with its most gorgeous pictures, before him

The study of natural history is the learning of the characters with which the wonderful story of nature is written, and I cannot conceive a more pleasing and natural introduction to its general study than entomology, of which I think the division Lepidoptera, the first portion of which will occupy the present volume, the most easy and attractive section The individual beauty of the insects in every stage, the ease with which they are preserved and the comparative facility with which a complete collection of British species may be formed, particularly of butterflies, of which we number scarcely more than eighty distinct species, render it a task of easy attainment, and carrying forth the student among trees and fields and flowers at the most delightful period of the year, its pursuit becomes very captivating, a most attractive first step towards the acquirement of a general knowledge of natural history

Many other temptations might be adduced to encourage the study of the natural history of butterflies, but I will only allude to one Botanists are sent at vast expense to every region of the earth to collect the most beautiful flowers of every clime for the decoration of our gardens, and the efforts of scientific gardeners are exerted to naturalise them to our climate, but though the plants of temperate regions have been found in many instances to bear the open air in our country, the gorgeous vegetation of the tropics can only be seen in England in the stove or the conservatory Yet, though we cannot transplant the flowers of the tropics to our bleaker soil, it appears by no means impossible that we may naturalise some of their splendid insects Their system of hybernation in the pupa case, in which state insects have been found to resist almost any degree of cold without injury, would shield them from the effects of our long winters Their development would not take place till the warmth of summer became sufficient, when we might see tropical butterflies flit from flower to flower, a splendid novelty to our gardens exhibiting colours more gorgeous than anything in the vegetable empire, and our lawns would be spangled with colours still more beautiful than those of the brightest flowers, and endowed, moreover, with the extra charm of motion If we cannot add the humming-birds of the Brazils to our garden luxuries, it seems probable that we might import many of the brilliant butterflies, in some instances even still more wonderful specimens of tropical colouring

These remarks may tempt some who have favourable opportunities to make the experiment; which with proper precaution in selecting such species as would find appropriate food in this country, (which would be easy, as several tropical plants are in most extensive cultivation here, of which it will suffice to name the potato and many species of Tropæolum), and also by selecting such as appear at the coolest seasons, might surely prove successful*. A few somewhat similar experiments have already been tried with complete success in transplanting English species from a part of the country where they were abundant, to one where they were comparatively unknown, and some few attempts have been made to naturalise continental species, but not upon a scale, or with the precaution, necessary for a fair trial I feel quite convinced that we might easily, for instance, render the beautiful P Podalirius, an extremely scarce if not absent species, as common as P Machaon

In conclusion, and more particularly addressed to those whose delicate sympathies shrink from making entomology a study on account of the *cruelty* of killing the necessary specimens, I may repeat a few of the arguments so often urged by eminent naturalists, but still little known except to the very small number who have made entomology a pursuit. Those who have not carefully examined the subject have got a few texts from poets or moralists, which serve them on all occasions as argument on the side of the alleged cruelty But they should recollect that poets are not always good naturalists Milton, for instance, speaks of the leaves of the banyan as " broad as an amazonian targe " an assertion quite at variance with fact On the other hand, men of science make sometimes but poor critics of art, like a celebrated French philosopher of the period of Louis XIV, who speaks of Michael Angelo and Raphael as "those *Mignards* of their age " Great men of all vocations may thus occasionally make mistakes in speaking of subjects which they have not carefully studied, and even Shakspeare shows that he was no entomological anatomist, when he asserts that the insect we unheedingly tread upon

> " In corporal suffering finds a pang as great
> As when a giant dies † "

Minute dissections, and the closest anatomical examinations, have proved that though insects are

---

* It should be tried both with eggs and pupæ

† I quote this passage in its original and long-accepted meaning, although several writers upon the subject have endeavoured to twist it to a different sense (Bennett, Zoological Journal, No 18, p 19, Mr Bird, Entomological Mag,) and make it appear that Shakspeare not only made no mistake, but on the contrary understood the physiology of insects as well, at a time when entomology was unknown as a science, as the men who during the last century have devoted their entire lives to the study of the subject by means of the most minute dissections and the most careful analyses

The passage occurs in 'Measure for Measure,' when Isabella encourages the condemned Claudio to suffer death rather than dishonour

> " Darest thou die ?
> The sense of death is most in apprehension,
> And the poor beetle, that we tread upon
> In corporal sufferance feels a pang as great
> As when a giant dies "

Mr Bird asserts that Shakspeare's intention was to show "that death was most in apprehension, and that even a beetle, *which feels so little*, feels as much as a giant " If Shakspeare, who is never mystical or indistinct, and had not the means of knowing that beetles felt little, had meant this, his passage would have been to the effect, that

> " The dying giant,
> In corporal sufferance, feels no greater pang
> Than the poor beetle that we tread upon "

The true meaning is very evident Shakspeare wishes to convey that a man of courage ought to be able to look calmly upon death as the common lot of every living thing—a pang to be borne at every moment by creatures unprovided with the moral strength of man that at every possessed of nerves, they have no well-defined organ representing our brain, the organ of concentrated feeling, where all the nervous conductors meet They have instead, a chain of ganglia or bundles of nervous substance, from each of which, nerves branch out to the contiguous parts, so that the sensations are not all carried to one grand central focus of acute sensibility as with us, but form, as it were, separate systems, any of which might be destroyed without communicating its sensation to the rest

Mr Haworth, the well-known English entomologist, being in a garden with a friend who firmly believed in the acute susceptibility of insects, struck down a large dragon-fly, and in so doing accidentally severed its long abdomen from the rest of its body. The mutilated insect, after this misfortune, felt so little inconvenience or loss of appetite, that it greedily devoured two small flies Mr H then contrived to form a false abdomen, to create such a balance to the rest of the body as would enable it to fly, after which it devoured another insect, and on being set at liberty, flew away with the greatest glee as if it had received no injury It is well known that large moths found asleep in the day-time may be pinned to the trunks of trees without their suffering a sufficient degree of pain to awake them, and only at the approach of twilight do they seek to free themselves from what they must doubtless consider an inconvenient situation * Many other equally striking facts might be adduced to prove that *cruelty* is not an objection to be made to the practical study of entomology Besides, cruelty is an unnecessary infliction of suffering, as when a person is fond of torturing creatures from mere wantonness without any useful end in view, or has recourse to circuitous modes of killing, where direct ones would answer equally well † The sportsman may perhaps be said to be cruel, for his primary object is amusement and unlike the entomologist, he is not adding to the general stock of knowledge But all dispute may be ended by a very simple expedient, for by dipping the entomological pin in prussic acid previous to piercing the insect, the effect is instantaneous

In the ensuing plates, the caterpillar, chrysalis, and imago, of every British species of butterfly will be given as far as they are known, and amateur collectors are invited to preserve, and make careful drawings of every caterpillar they may find which is not figured in this work, carefully supplying them with proper food till the change to the pupa state By such a course, pursued simultaneously by many individuals in different parts of the country, the gaps in the natural history of British Lepidoptera may be rapidly filled up, to which desirable object I trust the present work may conduce H N H

At the end of the volume will be given ample directions for rearing caterpillars from the egg, and for setting out and preserving the perfect insects

---

incautious step we may cause that pang of death to be borne by some poor insect which feels as keenly as ourselves, though deprived of the means of expressing pain or the mental strength to overcome it and that therefore man, the most perfect of created beings, ought not with his superior advantages to shrink from the pang which the poor beetle bears, when honour bids him face it This is very philosophically reasoned by Isabella, and is a strong argument to induce her brother to meet death rather than disgrace, the simile of the beetle being only rendered inappropriate by modern discoveries in entomology which have proved that a crushed beetle cannot feel "a pang as great as when a giant dies," but on the contrary that all insects are, from the nature of their structure incapable of acute sensations of pain H N H

* Vide Knowledge for the People (Zoology), and Encyclop Brit, art Animal kingdom

† Vide Kirby and Spence, Introduction to Entomology

BRITISH BUTTERFLIES

AND

THEIR TRANSFORMATIONS.

ORDER LEPIDOPTERA

THE beautiful tribes of Butterflies and Moths constitute one of the primary divisions or orders of winged insects, and are easily distinguished by several characters not found in any other annulose animals

The wings, four in number, are of a membranous texture, covered on both sides with innumerable minute scales resting upon each other like the tiles of a roof, and easily removed It is to these scales that the insects are indebted for their splendid colours, the membrane of the wing itself being colourless

The head is free,—that is, not received in a frontal cavity of the thorax,—and is furnished on each side with a large compound eye, and above with a pair of elongated antennæ, variable in form, not only in the different species, but also often in the sexes of the same species, and which in the butterflies are almost always terminated by a club The mouth occupies the lower part of the face, and appears at first sight to consist only of a long tongue, which the insect folds and unfolds in a spiral manner at will, and of a pair of scaly or hairy appendages, serving as a defence to the spiral apparatus when coiled up, but a more minute examination shows that the mouth is much more complicated in its structure, and that it exhibits all the parts (although sometimes in a rudimental state) of the mouth of the biting insects In fact, by denuding the front of the head of its scales, two minute triangular pieces are observed at a small distance apart above the origin of the spiral instrument, and which are the rudimental mandibles, here apparently useless, as is also the small conical upper lip placed between these two rudimental jaws, below which on each side is an oval plate soldered to the head, from the upper part of which arises one of the lateral halves of the spiral part, which in effect is composed of the two lower jaws extraordinarily elongated, and applied together so as to form a sucking tube, at the base of each portion of this tube is a minute tubercle, which in some species is developed into an elongated pair of feelers, or maxillary palpi, the labial palpi being the large feelers between which the spiral maxillæ are placed when at rest, and arising from the sides of the lower lip, which, like the basal part of the maxillæ, is soldered to the head

The transformations of these insects, which have attracted the attention of the most incurious observer from the earliest period also serve to distinguish them from all other insects The females deposit a considerable number of eggs, from which are hatched small worm-like jointed animals, furnished with a scaly head armed with a mouth and powerful jaws, six short scaly legs, attached in pairs to the three segments succeeding the head, and a variable number of short thick fleshy legs attached in pairs to the posterior segments of the body

The appearance of these larvæ is extremely variable, some being smooth, others warty, some hairy, &c. Their food consists almost entirely of vegetable matter. Whilst in this state they cast their skins several times, and when full-grown this operation is again repeated; but instead of the insect appearing as a caterpillar, it now more nearly resembles an Egyptian mummy, on minutely examining which, however, we can trace the rudiments of most of the limbs of the perfect insect, but closely applied to the body and covered by a general slender pellicle; the future wings occupying the sides of the anterior part of the body, between which are to be observed the leg-cases and the antenna-cases. The form of these chrysalides, aureliæ, or pupæ (as the insects are termed in this state), varies greatly: those of butterflies may almost always, however, be distinguished by having several angular prominences in various parts of the body, whilst those of moths are conical and not angularly tubercled. This peculiarity seems dependent on the circumstance that the caterpillars of the latter tribes enclose themselves in cocoons or cases entirely of silk, or of silk mixed with various extraneous materials, in which angular prominences on the body would be inconvenient to the inclosed insect: the caterpillars of butterflies, on the contrary, rarely form cocoons, but are transformed to pupæ in the open air. After remaining a certain period in this state, the time for the bursting forth of the perfect insect arrives; and after slitting the pupa skin in several directions, it disengages itself from its exuviæ, gradually extends its wings, and assumes all the beautiful characteristics of its perfect state.

> "Behold, ye pilgrims of this earth, behold!
> See all, but man, with unearn'd pleasure gay,
> See her bright robes the butterfly unfold,
> Broke from her wintry tomb in prime of May!
> What youthful bride can equal her array?
> Who can with her for easy pleasure vie?
> From mead to mead with gentle wing to stray,
> From flower to flower on balmy gales to fly,
> Is all she hath to do beneath the radiant sky."
>
> THOMSON'S *Castle of Indolence*.

The Lepidopterous insects were divided by Linnæus, the great classifier of the animal and vegetable kingdoms, into three primary genera, Papilio, Sphinx, and Phalæna, each subdivided into minor groups, and corresponding with the butterflies, hawk-moths and moths of English collectors. As, however, the number of species became more and more extended, and a more minute investigation of the characters of the species was made, it became necessary to introduce a much more extended mode of distribution, whereby the order was divided into three principal sections, Diurna, Crepuscularia, and Nocturna (corresponding with the three Linnæan genera). These have been again subdivided into families, and the latter into numerous genera and subgenera. The Crepuscularia and Nocturna, or the hawk-moths and moths are, however, much more closely allied together than either of them are to the Diurna; so that M. Boisduval, the most recent author upon the order, has proposed to adopt only two principal sections, Rhopalocera, or those with clubbed antennæ (butterflies), and Heterocera, or those with antennæ of variable shape but never clubbed (hawk-moths and moths). This arrangement appearing to me the most natural of any yet published, I have adopted it in my "Introduction to the Modern Classification of Insects;" but in the present work it will be convenient to adopt the previous arrangement of Latreille, the butterflies and hawk-moths being intended to form the subjects of the present volume, whilst the moths, or the section Nocturna, will form a subsequent volume.

SECTION I

LEPIDOPTERA DIURNA

The Diurnal Lepidoptera, or butterflies corresponding with the Linnæan genus Papilio, are distinguished not only by having the antennæ long and slender, and terminated in a larger or smaller club, and in the terminal family hooked at the tip, but also by the want of a bristle at the base of the anterior margin of the hind wings beneath, which, passing through a loop on the under side of the fore-wings of the moths, retains them in their proper place during flight The wings, moreover, when at rest, are mostly carried erect over the back, their upper surfaces being brought into contact The flight of these insects is diurnal Their caterpillars are constantly furnished with sixteen feet (six thoracic, eight ventral, and two anal) Their chrysalides are almost always naked, attached by the tail, and often by a girth round the middle of the body; they are often angular in their form, and are scarcely ever enclosed in a cocoon

The Diurnal Lepidoptera are divisible into the six following families —1 Papilionidæ (including two sub-families, Papilionides and Pierides) 2 Heliconidæ (including the Danaides) 3 Nymphalidæ (including the Hipparchides or the Satyrides of Boisduval, or Thysanuromorpha of Horsfield, and some other minor tribes separated by Boisduval) 4 Erycinidæ 5 Lycænidæ 6 Hesperidæ The last family differs from all the others in the habit of the caterpillars rolling up leaves, within which they undergo their transformation

---

FAMILY I

PAPILIONIDÆ, Leach.

This family consists of some of the most gigantic species of butterflies distinguished by having all the six feet in the perfect state fitted for walking, the anterior pair not being more or less rudimental; the hind tibiæ have only a single pair of spurs at the tip, the tarsal ungues are distinct and exposed, the antennæ are never hooked at the tip, the club being distinct, but variable in form, the palpi are variable, but the third joint is never suddenly slenderer than the rest and naked, the central cell of the hind-wings is always closed behind by a nerve The caterpillars are elongated, nearly cylindrical, and naked. the chrysalides are attached not only by the ordinary anal hooks, but also by a girth round the middle of the body In one genus (Parnassius) it is, however, inclosed in a rough cocoon

This family is divisible into two sub families, Papilionides and Pierides The first sub-family, Papilionides, has the anal edge of the hind-wings concave, or cut out to receive the abdomen, the anterior tibiæ have a spur in the middle, and the tarsal ungues are simple The caterpillars are furnished with two fleshy retractile tentacles, forming a fork upon the back of the segment succeeding the head

GENUS I

PAPILIO, Linnæus

The antennæ are elongate, with a moderate-sized club, gradually formed and somewhat curved, the palpi are very short, so as to be scarcely seen,—they appear only two-jointed, the third joint being almost obsolete, the spiral tongue is long, the eyes are large and naked, the abdomen rather short, and ovate-conical, the wings are elongate, and more or less toothed at the edges, the posterior pair being often produced into a long tail, and having the anal margin cut out to allow the motion of the abdomen; the strong central nerve of the fore-wings emits four branches behind, and the middle cell of the hind-wings is closed and emits six nerves The fore-legs are alike in both sexes, the two fore-legs being fitted, as well as the four hind ones, for walking, the anterior tibiæ have a single strong spur at the middle, the four hind tibiæ have two long spurs at the tip of each The larvæ are naked, and furnished on the neck with a fleshy furcate tentacle which they are able to retract or exsert at will The chrysalides are attached by the tail, and girt round the middle with a silken thread, with the head pointing upwards, and forked, or bimucronate

This genus is extremely numerous, Boisduval having described as many as two hundred and twenty-four species, exclusive of several which he has detached under other generic names They are mostly of large size, and are found in the Tropics, only three or four species being natives of Europe

DESCRIPTION OF PLATE I

Insects —Fig 1 Papilio Machaon (the swallow tail Butterfly) 2 The Caterpillar 3 The Chrysalis
" Fig 4 Papilio Podalirius (the scarce swallow tail B) 5 The Caterpillar 6 The Chrysalis
" Fig 7 Gonapteryx Rhamni (the brimstone B ,) the male. 8 The female 9 The Caterpillar 10 The Chrysalis
Plants —Fig 11 Daucus carota (wild Carrot) 12 Prunus spinosa (Sloe) 13 Rhamnus Catharticus (Buckthorn)

I have here introduced Papilio Podalirius, though still considered doubtful by some entomologists, as a British species My figure is drawn from an Italian specimen of my own taken near Tivoli, which differs from the individual figured by Mr Curtis, in having the black stripes less strongly marked * Papilio Machaon and Gonapteryx Rhamni are from English specimens in my own collection, and I should have liked to have added G Cleopatra from a beautiful Italian specimen, as in most countries of Europe it is equally common with G Rhamni, but no instance of its being seen in England has, I believe, yet occurred † It might however, I should think, be easily imported, but Boisduval asserts that G Cleopatra and G Rhamni have both been reared from the same eggs, and that the caterpillars offer no perceptible difference even in the markings, thus inferring that the more rich colouring of G Cleopatra is attributable to the effect of climate upon the more robust individuals It is singular, however, if this be the case, that similar varieties do not occur in favourable climates in other parts of the world where G Rhamni is common, in Nepaul for instance, or parts of North America, where the slight modifications effected in G Rhamni by climate or food are but slight H N H

---

* The unusual darkness of the markings of Mr Read s English specimen of P Podalirius, tends to confirm the really indigenous character of that specimen, which seems to be completely supported by the fact that the individuals taken further south than the one figured in Plate I, namely in Spain and the north of Africa, have the ground colour of the wings " toujours sensiblement plus blanc," which has induced M Godart to regard them as forming a distinct species under the name of P Feisthamelii J O W

† The variety figured by Mr Curtis cannot be considered even an approach to G Cleopatra H N H

SPECIES 1.—PAPILIO MACHAON. THE SWALLOW-TAIL BUTTERFLY.

Plate 1, fig. 1—3.

SYNONYMES.—*Papilio (Equites Achivi) Machaon*, Linn. Syst. Nat. ii. 750. Donovan Brit. Ins. 6. pl. 211. Lewin Brit. Butt. pl. 31. Harris Aurelian, pl. 36. Westwood Ent. Text Book, tab. 1. fig. inf. ditto Introd. ii. p. 332, fig. 95. 1—10. Duncan Brit. Butt. pl. 4, fig. 1.

Papilio Reginæ, De Geer. Gen. 6. 30. 5.
Iasonides Machaon, Hubner.
Amaryssus Machaon, Dalman Pap. Suec. 85. 1.

This beautiful butterfly varies from three inches to three inches and nearly three-quarters in the expanse of the wings, which are of a yellow colour with black markings, the fore wings having a large patch of black at the base; the anterior margin is black, with three large black subcostal marks; the nerves are also broadly black, as is also the apical margin, in which are eight yellow lunules, above which is a thick sprinkling of minute yellow scales. The posterior wings are more strongly denticulate at the edges, and produced behind into a pair of rather long tails; they are yellow, with the inner margin and a very broad apical border black, the latter with six yellow lunules, above which is a thick sprinkling of blue scales. The anal angle is ornamented with a brick-red eye, margined with yellow beneath and with blue above, the latter having also a black crescent above it.

The under side of the wings is much paler than the upper, the black markings being less extended; the apical yellow lunules of the upper sides are replaced by a narrow continuous bar, above which the yellow irroration is much stronger; the broad black apical bar of the hind wings is much paler, the black being confined to the curved margins of the bar, and in the middle of the hind wings are three triangular brick-red spots; a spot of the same colour also exists in the squarish yellow submarginal spot nearest the fore edge of the hind wings.

This species is very widely dispersed, being found all over Europe, Siberia, Syria, Egypt, the coast of Barbary, Nepaul, Cachemere, and the Himalayan mountains, from which last locality I possess a specimen captured by Professor Royle, which scarcely exhibits the slightest differences when compared with English specimens. In our own country it chiefly occurs in the fenny districts of Cambridge and Huntingdonshire, but it has also been captured in Hampshire, Middlesex, Sussex, Essex, and Kent. The caterpillar is of a fine green colour, with velvety-black rings, spotted alternately with fulvous-red. It is found in June and September, there being two broods in the year according to Boisduval; but this is doubted by Stephens, the perfect insect being taken from the beginning of May to the end of August in England. It feeds upon various umbelliferous plants, especially on the marsh-milk parsley (Selinum palustre), fennel (Anethum fœniculum), and wild carrot (Daucus carota), preferring the flowers. The fork-like tentacle on the neck is of a red colour, and emits a strongly-scented liquor when alarmed, by which it is said to drive off the Ichneumon flies. The mode in which the transformation of this butterfly is effected has been carefully investigated by Reaumur. When full-grown, the caterpillar discharges from the spinning apparatus in the middle of the under part of the mouth a small quantity of silk, forming it into a little mass which it lays hold of with its hind pair of feet; it then attaches another thread on one side of the twig at some distance in advance of the small mound, and gradually forms it into a loop, attaching the other end of the thread to the opposite side of the twig, and holding it open by means of its fore legs; it then spins a sufficient number of similar threads, until the loop has acquired sufficient strength for its destined use. When it is completed, the insect still holding it open by means of its fore legs, somewhat in the same way as a skein of silk is held on the hands whilst being wound off, slips its head between these legs, and thus passes the loop over its back, and by the repeated action of the anterior segments, it gradually brings it to that part of the body best calculated to balance it when it shall have assumed the chrysalis state. It sometimes however happens, that notwithstanding all its care, the threads of the loop slip off its legs. This is indeed a woful calamity to the poor larva, which has the greatest difficulty, and is sometimes unable, to collect the threads of the loop upon its legs, trying every contortion of limb to effect this purpose, but sometimes in vain. Should it however succeed, the body is stretched forward in a right line, and remains in this position until the skin is cast, being slit down the back by the contortions and annular contractions of the insect, the girth being too loose to form a material hindrance to its being slipped backwards to the tail, where it is ultimately thrown entirely off.

SPECIES 2.—PAPILIO PODALIRIUS. THE SCARCE SWALLOW-TAIL BUTTERFLY.

Plate fig. 4—6.

SYNONYMES.—*Papilio (Equites, Achivi) Podalirius*, Linn. Syst. Nat. ii. 751. Donovan Brit. Ins. 4, pl. 109. Lewin Brit. Butt. pl. 35. Curtis Brit. Entomol. pl. 378. Westwood Ent. Text Book, pl. 4, fig. 1. Duncan Brit. Butt. pl. 4, fig. 2.

Podalirius Europæus, Swainson.
Iphiclides Podalirius, Hubner.

This fine butterfly is about the same size as, or rather larger than, P. Machaon, varying from three inches to four inches and a half in expanse. The hind wings are also longer; the ground colour of the wings is much lighter, being straw or cream yellow. The fore wings have two broad black bars crossing them near the base, and extending half-way across the hind wings; the second of these is succeeded by a black bar extending from the costa to the middle nerve of the wing, and this by a stripe entirely crossing the wing, but gradually narrowed behind; this is accompanied by another shorter narrower bar extending from the fore margin nearly half across the wing; the apex of the wing has a broad black edge narrowed towards the posterior angle, and divided by a narrow yellow stripe. The extremity of the hind wings is black, with a black anal patch, having a blue lunule in its middle, and bordered above with red, forming an eye. The posterior margin has four or five lunules of blue specks, and the edges are indented with yellow crescents; the tip of the tails is also yellow.

The under side is paler than the upper, the black markings being less extended, and the fascia on the middle of the hind wings is formed of two narrow black lines, the outer one being edged within with orange.

Like P. Machaon, this species inhabits the whole of the Southern and temperate parts of Europe (being found plentifully near Berlin, at Moscow, and even at Hamburgh), the North of Africa, and Asia Minor. Its claim, however, to be ranked as a British species has occasioned much controversy. The great Ray says that he had found it "*nè male memini* in Anglia," and Berkenhout, in his Outlines of British Natural History, that it is "rare in woods." Dr. Abbott stated to Mr. Haworth, that he took a specimen near Clapham Park Wood in Bedfordshire. The Rev. F. W. Hope, F.R.S., has informed me that he took a specimen near Netley; and W. H. R. Read, Esq. has also informed me that whilst at Eton, about the year 1826 (the last year of his being at that school), he took a specimen on the wing between Slough and Datchet, previously to the month of July, when the vacation commences. This specimen, which is a very dark-coloured one, has been figured by Mr. Curtis, who gives a different date to the capture from that mentioned to me by Mr. Read himself. In my copy of Haworth's Lepidoptera Britannica, which belonged to Donovan, I find the following pencil note in the hand writing of the latter, opposite P. Podalirius:—"One in Mr. Swainson's cabinet, which he told me was taken by his brother-in-law, Captain Bray, I think he said in the Isle of Wight."

The caterpillar is short and thick, especially towards the head, becoming more taper towards the tail, it is green, but varying to yellowish red, with a slender yellowish dorsal stripe, and a lateral one above the legs, on each side are also oblique yellowish lines dotted with red, the head is small, and the neck is furnished with a forked tentacle It feeds on the apple, peach, almond, plum, sloe, and especially on the sloe-thorn, and not on the species of Brassica, as stated by Fabricius and Stephens

---

GENUS II

PARNASSIUS, Latr DORITIS, Fabr Stephens

The antennæ are short, terminated by a straight and nearly oval club, the palpi are longer than the head, above which they are slightly elevated, they are composed of three distinct joints of nearly equal size The body is robust and hairy, the wings parchment-like and sparingly clothed with scales, and of a somewhat oval form, with the edges entire, the hind pair has the anal edge entirely excised so as to allow the free action of the abdomen

The caterpillars are thick, cylindrical, covered with little setose tubercles, the segment succeeding the head being furnished with a fleshy fork-like appendage similar to that of the genus Papilio The chrysalis is cylindric-conical, covered with a bluish powder, enveloped between the leaves in a slight silken web, and supported by several transverse threads

This is an exceedingly interesting genus, beautifully connecting the Papilionides with the Pierides In the general appearance of the butterfly, the three-jointed palpi and short antennæ, it approaches the latter, but in the characters of the larva, and especially in the curious retractile fork of the neck, and the excised hind-wings of the perfect insect it resembles the genus Papilio The chrysalis, with the exception of the head, resembles those of the moths of the genus Catocala, especially in the blue powder with which it is covered, whilst the mode in which it is inclosed in a kind of cocoon formed of leaves loosely fastened together gives it some relation with the Hesperidæ The females are further remarkable for possessing a small corneous pouch at the extremity of the abdomen, which is found in no other Lepidopterous insect

The genus Parnassius was first proposed by Latreille in his Histoire Naturelle, with P Apollo for its type, the name of Doritis not having been published until several years subsequently, with the same type I have accordingly followed M Boisduval in restoring the former name to this genus, although he appears to me to have erred in giving the name of Doritis to other insects with which Fabricius was totally unacquainted It would certainly have been preferable to have proposed a new generic name for these last butterflies

The species of Parnassius are few in number, and chiefly found in the Alpine regions of the old world The larvæ feed solitarily upon the species of Sedum and Saxifrage growing in such situations

SPECIES 1.—PARNASSIUS APOLLO. THE CRIMSON-RINGED BUTTERFLY.

Plate vii.

SYNONYMES.—*Papilio Apollo*, Linn. Syst. Nat. ii. 754. Donovan Brit. Ins. vol. xiii. pl. 433. Haworth Ent. Trans. i. p. 332. | *Doritis Apollo*, Fabricius. Stephens, Duncan Brit. Butt. pl. 11, fig. 1. *Parnassius Apollo*, Boisduval Hist. Nat. Lep. i. 395.

The expansion of the wings of this fine butterfly varies from three to three and a half inches. The wings are white, the base and fore margin of the anterior being dotted with black, the apex of the fore wings being transparent, and preceded by a transverse sinuous row of blackish dots; the disc of the fore wings is also marked with five black spots; the posterior wings have two large eye-like spots on the disc of a red colour, surrounded with black, the centre being white; there are also two other smaller black spots often united together at the anal angle of the hind-wings. The under side of the fore wings differs in having the innermost black spot, and sometimes also the apical one, often marked with red in the middle, whilst the hind wings differ in having four red spots at the base edged with black; the two spots also at the anal angle of these wings is also red, edged with black, with a white middle.

The caterpillar is of a velvety black colour pubescent, with orange-coloured dots, and small blue tubercles; the neck is furnished with a yellow forked appendage, capable of being entirely withdrawn into the segment. It feeds on Saxifrage and Crassulaceæ.

The butterfly is common in most of the alpine regions of Europe; but it appears to be very questionable how far it is truly a British species, the original specimen, supposed to have been captured in one of the Hebrides, having been received in a box of insects from Norway. Mr. Duncan however, states that he had been assured that it had been noticed on the wing in some part of the west coast of Scotland in the summer of 1834.

The second sub-family of the Papilionidæ, PIERIDES, has the anal edge of the hind wings formed into a gutter to receive the abdomen; the anterior tibiæ do not possess a spur in the middle; and the tarsal ungues are one or two-dentate.

The caterpillars are not furnished with a nuchal fork. They are slightly pubescent, and rather slender at each end of the body.

GENUS III.

GONIAPTERYX, WESTW. GONEPTERYX, LEACH.

The antennæ are rather short and robust, terminated by a club gradually formed, commencing nearly at the middle of the antennæ, the apex not compressed and slightly truncate; the scales on the front of the head form an erect tuft; the palpi are as long as the head, distinctly three-jointed, the third joint small; the wings are ample, the anterior angulated at the tip and the posterior nearly in the middle of the hind margin; the fore legs are alike in both sexes, the tibiæ not armed with a spur in the middle, and the tarsal ungues bifid, with slender pulvilli The larvæ are elongated, slightly pubescent, attenuated at each end The chrysalis is gibbose, much bent, terminated like a spindle at each extremity, always attached by the tail and by a transverse girth across the middle

M Boisduval rejects Dr Leach's name for this genus, Gonepteryx (misquoted by the former under the name Gonopteryx), because it is too much like Gonoptera, proposed *long afterwards* by Latreille for another genus of Lepidoptera, and because names ending in *ptera* have but little euphony, and ought only to be used in Ichthyology, where they are more prevalent All these reasons appear to me insufficient I have, therefore retained Dr Leach's name with a slight alteration, making it more in accordance with its Greek derivatives γωνια an angle, and πτερον a wing

---

SPECIES 1—GONIAPTERYX RHAMNI THE BRIMSTONE BUTTERFLY

Plate i fig 7—10

SYNONYMES —*Papilio (Dan Cand) Rhamni*, Linn Syst Nat ii p 765 Donovan Brit Ins 5 pl 145 Lewin Brit Butt pl 31 Albin Brit Ins pl 3, fig 3 e h
Gonepteryx Rhamni, Leach Stephens Curtis, Duncan, Brit Butt pl 5, fig 1
Goniapteryx Rhamni, Westw Introd Gen Syn p 87
Rhodocera Rhamni, Boisduval Hist Nat Lep 1 p 602
Anteos Rhamni, Hubner
Ganoris Rhamni, Dalman

This butterfly varies in expanse from two inches to three inches and a half The male has the upper side entirely sulphur-yellow, and the female greenish white, with an orange spot at the extremity of the discoidal cell of each wing, and some very minute ferruginous points at the place of union of the nerves with the margins of the wings The under side of the wings is paler than the upper, especially in the males, and the orange discoidal spot is replaced by a ferruginous dot, whitish in the centre, between which and the marginal ferruginous points is a row of fuscous spots

Mr Curtis has figured a variety of this insect captured near Peckham, with the upper wings variegated with orange, slightly as in G Cleopatra, thus proving the correctness of the statement made to me by M Boisduval, that he had reared G Rhamni and Cleopatra from eggs deposited by the female of the former, the larvæ producing the latter offering no variation from those from which the latter were reared (See also his Hist Nat Lepid i p 602)

To no species of butterfly can we apply with greater effect, but with a little alteration, Mrs Barbauld's beautiful image of the origin of the snowdrop —

> As if "Floras breath, by some transforming power,
> Had changed" a flower into a butterfly

Sporting about in some flowery nook in the very first sunny days of February and March, this butterfly looks more like the petals of the primrose over which it hovers, floating on the breeze, than a living creature These early specimens have survived the winter, and produce eggs from which a fresh brood of butterflies is produced in May, and another in the autumn, some of which last again survive the winter

The caterpillar is green, finely shagreened with black scale-like dots on the back, with a whitish or pale-green line on each side, the upper edge of which is shaded off into the general colour It feeds on the buckthorn (Rhamnus catharticus), and the berry-bearing alder (Rhamnus frangulus), as well as on Rhamnus alaternus The chrysalis is green, with several reddish dots, it is very gibbous in the middle, and attenuated like the end of a boat in front, it is attached by the tail on a perpendicular branch, and fastened with a loose silken thread round the middle of the body, the pupa state lasts about a fortnight This butterfly occurs commonly in various parts of England, as far north as York, Windermere, and Newcastle, but Mr Duncan states that it has not yet been found in Scotland

---

GENUS IV

COLIAS, Fabricius

The species of this genus, like the preceding, are distinguished by the brilliant yellow or orange colour of their wings, but they are more or less bordered at the tips with black, and are never angulated The fore wings exhibit also on both sides a discoidal black spot, and the posterior a central spot, which is orange above and generally silvery beneath The antennæ are short, nearly straight, gradually clavate to the tip, which is truncate, the palpi are shorter than in Gonepteryx, the head has no frontal tuft, the fore wings are sub-triangular, and the posterior are rounded, the fore-legs are alike in both sexes, the tarsal ungues bifid, and the pulvilli very minute (A highly magnified figure of the ungues and their appendages is given in the Crochard Edition of the Règne Animal, Insectes, pl 132, fig 3, c) The caterpillar is naked, elongate, cylindric, very finely setose and tubercled, the chrysalis rather short, subangulated, gibbous, slightly beaked in front, attached by the tail and by a girth behind the thorax

From the great similarity of some of the numerous species of this genus and their apparent variation, much confusion has occurred in the investigation of the British species, Mr Stephens describing four (exclusive of P Palæno, Linn, a reputed British species, and P Helice, Haw, a presumed variety of C Edusa), whilst Mr Curtis only admits two The Rev W Bree has published a memoir on the British species in the Magazine of Natural History, No 26

DESCRIPTION OF PLATE II.

Insects —Fig 1 Colias Edusa (the clouded-yellow Butterfly), the male 2 The female 3 The Caterpillar 4 The Chrysalis

" Fig 5 Colias Hyale (the pale clouded-yellow B,) the male 6 The female 7 The Caterpillar

" Fig. 8 The pale female variety of C Edusa, considered by some authors as a distinct species under the name of C Helice

Plants —Fig 9 Sylibum Marianum (Milk-thistle) 10. Festuca gigantea (Giant fescue grass)

In this plate I have given both male and female of two of our most brightly-coloured native butterflies, and have been careful, by placing C Edusa and C Hyale on the same plate, to show the marked and striking difference of the two species, the former being of a rich orange-colour, whilst the latter is of a pale sulphur; notwithstanding which, by bad colouring and other defects, they have been completely confused in many other works They are all from fine specimens, particularly the female Edusa, from the collection of Mr Westwood The C Helice, or pale female variety of C Edusa, is from an Italian specimen in my own collection, but it differs in no respect from the pale English varieties H N H

SPECIES 1.—COLIAS EDUSA. THE CLOUDED-YELLOW BUTTERFLY.

Plate 2, fig. 1, 2, 3, 4, and 8.

SYNONYMES.—*Papilio Edusa*, Fabricius Ent. Syst. v. iii. part 1, p. 206. Donovan Brit. Ins. 7, pl. 238, fig. 2 (female). Harris Aurelian, pl. 29, fig. ii. ♂, fig. iii. ♀.
Colias Edusa, Stephens, Curtis, Duncan Brit. Butt. pl. 5, fig. 2.
Papilio Hyale Esper. Donovan Brit. Ins. 2, pl. 43, fig. sup. ♂.
Papilio Electra, Lewin, pl. 32 (and Linn. Syst. Nat. ii. 764 teste Newman Ent. Mag. 1, 85).

The expansion of the wings of this species varies from nearly two inches to two inches and a half. The upper surface of the disc of the wings in both sexes is a rich orange colour, the males having a round discoidal black spot on the fore wings, and a broad black apical margin irregularly toothed within, extending through both wings, with several narrow orange lines running through the black border, indicating the place of the nerves; the disc of the hind wings is somewhat darker, with a large discoidal brighter-coloured orange patch. The upper side of the female differs in having the broad apical border marked with several irregular yellow spots, and more indistinctly indicated in the hind wings, which are darker and yellower than in the males.

Beneath, both sexes are nearly alike, the disk of the fore-wings being lighter orange, with a black discoidal spot, the margins greenish, with a row of blackish spots at some distance from the apical margin; the hind wings are greenish, with a round silver discoidal spot surrounded with red, and accompanied by a smaller silvery dot; between this and the apical margin is a row of brownish red dots. Most modern Entomologists are agreed in regarding the Papilio Helice of Hubner and Haworth, figured by Stephens (Illustr. Haust. pl. 2*, and our fig. 8), as a variety of the female of Colias Edusa, from which it differs in having the ground colour of the disc of the wings, as well as the spots in the black apical margin, yellowish white. No corresponding variety of the male has yet been observed. The insect figured in the next plate, under the questionable name of Colias Chrysotheme, has also been regarded by Mr. Curtis as a variety of C. Edusa.

The caterpillar of C. Edusa, which feeds upon Medicago lupulina, various species of Trifolium, and other leguminous plants, is green, with a lateral stripe varied with white and yellow, and with an orange dot on each segment. The chrysalis is green, with a lateral yellow line and several ferruginous dots.

Boisduval gives Europe, Egypt, the coast of Barbary, Nepaul, Cachemere, Siberia, and North America, as the localities of this species. Mr. Burchell is stated by Mr. Duncan to have found it in South Africa; but this I apprehend must have been the species described by Boisduval, from the Cape of Good Hope and Caffraria, under the name of C. Electra of Linnæus, by whom also the Cape of Good Hope was given as its locality. Hence from the similarity of the two species, it is that I have hesitated to consider our English species as the true C. Electra, as stated by Mr. Newman.

This is one of those species of butterflies whose periodical appearance (every three or four years, as stated by some writers) has so much perplexed Entomologists. Various opinions have indeed been suggested by authors, in order to account for this singular circumstance,—such as the failure of their natural enemies, the Ichneumonida, or insectivorous birds—an increased temperature—or the dormant state of the eggs until called forth by some latent coincidences. All these opinions are, however, but merely conjectural; nor can the matter be cleared up until a more minute inquiry into the habits of the species has been made.

SPECIES 2.—COLIAS HYALE. THE PALE-CLOUDED YELLOW BUTTERFLY.

Plate II. fig. 5—7.

SYNONYMES.—*Papilio (Danai Candidi) Hyale*, Linn. Syst. Nat. II. p. 764. Lewin Brit. Butt. pl. 33. Donovan Brit. Ins. 7. pl. 238, fig. 1 (male).
Colias Hyale, Ochsenheimer, Leach, Stephens, Curtis, Brit. Ent. pl. 242; Duncan Brit. Butt. pl. 6, fig. 1.
Le Soufre, Ernst. Pap. d'Europe.
Papilio Palæno, Esper. 1, pl. 4, fig. 2. Fischer Entomol. de la Russ. Lepid. pl. 11, fig. 1, 2 (Colias P.).

This species is from two inches to nearly two inches and a quarter in expanse. Its upper surface is of a sulphur colour in the males, or of a cream colour in the females; in other respects the sexes are nearly alike, having a black discoidal spot and an irregular, broad, black, apical margin, in which is an interrupted series of spots of the same colour as the ground of the wings; the hind wings are darker on the disc, with an orange-coloured discoidal spot, and the margin is very slightly and irregularly marked with black. The fore wings are beneath whitish yellow, with the apex orange yellow, having a row of transverse blackish marks parallel with, but at some distance from, the apical margin; the discoidal spot is black, with a yellowish middle; the hind wings beneath are orange yellow, with a large silvery spot, accompanied by a minute eye-like dot surrounded with reddish, and between these and the apex of the wings is a row of small blackish spots.

The caterpillar, which feeds on Medicago, various species of Trifolium, and other leguminose plants, is of a velvety green colour, with two lateral yellow stripes, and with black dots on the segments; the chrysalis is green with a yellow lateral line.

Although very abundant on the Continent (where it appears to be double-brooded, May and August, or September, being the times of its appearance), and extending to the north of Africa, Siberia, Cachemere, and Nepaul, this butterfly is much rarer in England than the preceding; the coasts of Kent, Sussex, and Suffolk, having furnished the greatest number of specimens. Others have been found in Epping Forest, and near Halvergate in Norfolk.

DESCRIPTION OF PLATE III.

INSECTS.—Fig. 1. Colias Chrysotheme? (small clouded yellow Butterfly), the male. 2. The female. 3. The underside.
" Fig. 4. Colias Europome? (the clouded sulphur B.). 5. Female. 6. The underside.
PLANTS.—Fig. 7. Lotus corniculatus (common bird's foot trefoil). 8. Lotus major (great bird's foot trefoil).

In this plate I have figured Colias Europome? and Colias Chrysotheme? C. Europome is evidently a distinct species, but is by many denied to be British; as, however, it is found in many English collections, this work would not be complete without it. It is from the Haworth specimens in the collection of Mr. Stephens. The caterpillar is unknown. C. Chrysotheme? is occasionally found, but is by many deemed a small variety of C. Edusa, and in Mr. Stephens's cabinet a small true Edusa is placed next to it, in which there is scarcely any perceptible difference. The discovery of the caterpillar, which is unknown, would solve the question. It should be sought for near the coast, where the trefoils are abundant, which are its probable food. I have placed it in the same plate with Europome to show the completely distinct colour of the latter species, Europome being of a decided citron, whilst Chrysotheme and Edusa are of a rich full orange. The colour of Europome is exactly midway between the pale sulphur of Hyale and the orange of Edusa, and perfectly distinct from both. I would therefore propose as English names, *clouded orange* for Edusa, *clouded sulphur* for Hyale, and *clouded citron* for Europome, as being more descriptive of the insects than those which, from being now in use, I have adopted above. H. N. H.

No. II.] [Price 2s. 6d.

BRITISH INSECTS AND THEIR TRANSFORMATIONS.

BRITISH BUTTERFLIES

AND

Their Transformations.

ARRANGED AND ILLUSTRATED IN A SERIES OF PLATES

BY H. N. HUMPHREYS, ESQ.

WITH CHARACTERS AND DESCRIPTIONS

BY J. O. WESTWOOD, ESQ., F.L.S.,

SEC. OF THE ENTOMOLOGICAL SOCIETY, ETC. ETC.

LONDON:
WILLIAM SMITH, 113, FLEET STREET.

MDCCCXL.

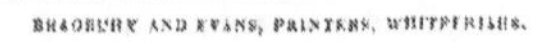

BRADBURY AND EVANS, PRINTERS, WHITEFRIARS.

SPECIES 3.—COLIAS EUROPOME? THE CLOUDED SULPHUR BUTTERFLY

Plate 3, fig 4—6

SYNONYMES.—*Papilio Europome*, Villers Ent 2, 17 19, and 4, 168, 19? Esper Schmetterl 1, pl 12, Suppl 18, fig 1, 2, and pl 100, cont 55, fig 5? Haworth Lep Brit p 13 Stephens Illust Haust pl 1 *

Papilio Palæno, Linn Syst Nat 2 764? Hubn Pap pl 86, fig 131, 135? Boisduval Hist Nat Lep 1 p 645

Eurymus Europome Swainson Zool Illus: n ser pl 70

In the opinion of the best modern Entomologists, Papilio Europome and Palæno are identical, the latter especially inhabits Sweden, as well as the Alps and Pyrenees Martyn, in his Aurelian's Vade-Mecum, introduced P Palæno as a British species, and Mr Haworth, in his Lepidoptera Britannica, P Europome, on the authority of the cabinets of Francillon and Swainson In the Butterfly Collector's Vade-Mecum, it is said to occur in meadows and roadsides near Ipswich, and in the Entomological Magazine, it is stated to have been noticed in the meadows near the confluence of the Avon and Severn, flying with great swiftness in August Its claim, however, to be regarded as indigenous, is still denied by several of our principal Entomologists, and Mr Curtis asserts that Mr Stephens' specimens from the collection of Francillon, mentioned above, and figured by him in his Illustrations, pl 1 * are identical with the North American C Philodice Mr Swainson also, as quoted above, has described and figured the individuals in his father's collection, which had been mended with the heads of Gonipteryx Rhamni, and which he states that he could not distinguish from C Philodice

As however there is a great resemblance between C Philodice and C Europome (or Palæno), and as the latter is a native of Sweden, and therefore not unlikely to occur in England, we have not thought it proper to reject it, especially since we have admitted Papilio Podalirius and Parnassius Apollo The following is a translation of M Boisduval's description of the Swedish species C. Palæno, or Europome, which it will be serviceable to compare with that of Mr Stephens, drawn from the asserted American specimens —

"Rather smaller than C Hyale Upper side of the wings of a slightly greenish yellow, with a rather broad black border, slightly sinuated on the inside, narrowed in the hind wings, and often not extending beyond the middle of the limb, the anterior having at the extremity of the discoidal cell a small blackish circle oblong, more or less marked, rarely wanting, or replaced by a blackish dot, the posterior wings with a small whitish discoidal spot The under side of the fore wings differs from the upper in having the margin replaced by greenish yellow or reddish colour, and in the small discoidal circle being more decided The under side of the hind wings is entirely of a reddish yellow, or greenish, with a silvery white discoidal spot, slightly circled with ferruginous The fringe of the four wings and the costa of the anterior beneath is bright rose-coloured The body is of a blackish yellow or greenish, with the prothorax rosy, the antennæ rosy, with the club darker coloured and yellow at the tip The female differs from the male in having the ground colour of the wings nearly white or yellowish-white on the upper side" The larva is described by Zetterstedt as pubescent, green, with yellow lines and black dots The pupa is unknown

---

SPECIES 4.—COLIAS CHRYSOTHEME? THE SMALL CLOUDED YELLOW BUTTERFLY

Plate 3 fig 1—3

SYNONYMES.—*Papilio Chrysotheme*, Esper, 1, pl 65, fig 3, 4?
Colias Chrysotheme, Stephens, Illustr Haust pl 2, fig 1, 2
Colias Edusa, var Curtis, Duncan

The insects regarded as this species by Stephens have been considered by several subsequent writers as small varieties of C. Edusa.* The figures indeed given of C. Chrysotheme, especially those of Boisduval (Icon. Hist. des Lépidopt. pl. 9, fig. 3, 4) differ materially from those of Stephens; and as C. Chrysotheme is described as a native of Hungary, Syria, and Southern Russia, it is most probable that that species is not a native of England. Boisduval has indeed noticed a character which will satisfactorily decide the specific identity of the English specimens with C. Chrysotheme or C. Edusa, the genus being divisible into two groups, C. Edusa belonging to the first, in which the males are provided with a glandular space or sac at the anterior edge of the hind wings near the base, whilst in the second group, to which C. Chrysotheme belongs, they are destitute of this sac.

The following is a translation of Boisduval's description of the true C. Chrysotheme, which I have here introduced in order that it may be compared with Mr. Stephens's description of the supposed English species:—

"Figure of C. Edusa, but about one fourth smaller. Upper side of the wings of a paler yellow, with the margin browner, divided in the fore wings by fine yellow nerves; the fore wings having moreover the costa broadly yellow. The discoidal spot is narrower, transverse, slightly marked, and edged with a little red. The under side of the fore wings nearly as in C. Edusa and the allied species, except that the discoidal spot of the fore wings has the centre rather pupilled with silver. The female is much paler than the female of C. Edusa, and the yellow orange colour only occupies the disk of the fore wings, and the yellow spots which divide the dark margin are larger, more marked, and of a much paler yellow colour."

The generic name Colias appears to have inappropriately derived by Fabricius from κολίας, a word used by the Greeks for some kind of fish.

Mr. Stephens gives Norfolk or Epping Forest as the locality of one of his British specimens.

GENUS V.

PIERIS†, Schrank. PONTIA, Stephens, &c.

In its present restricted state this genus consists of species, which from their common occurrence in our gardens throughout the summer, have attracted our earliest attention; their almost uniform white colour, and the places where they mostly frequent, having led to their receiving the ordinary name of Garden Whites. From the preceding genus they are distinguished by the more acute tip of the fore-wings, and by their longer and slenderer antennæ, which are terminated by a broad compressed and obtuse club. The palpi are short, three-jointed, nearly cylindrical, with the terminal joint as long as or rather longer than the second. The legs are long, slender, and alike in both sexes, the anterior pair being perfect.

The tarsi are terminated by two equal-sized hooklets much curved, each having a small tooth on its under side; between these hooklets is a long fleshy pulvillus, and each is laterally defended by a long conical hirsute appendage. The details of this curious structure are represented in the Crochard edition of the Règne Animal,

* It is to be observed that M. Boisduval describes no other variety of C. Edusa than the C. Helice. Can the small English specimens be C. Myrmidone?

† Derived from Πιερὶς, plural Πιερίδες, the Muses, a poetical licence similar to that used in giving the name Parnassus to the genus having P. Apollo for its type.

now in course of publication Curtis describes the claws as unidentate or bifid, but I have found their structure uniform in all the species I have examined, agreeing also in this respect with the black-veined white and orange-tipped butterflies

The wings are opake, and thickly clothed with scales, thus disagreeing with the black-veined white and Apollo The upper wings are at once distinguished from those of all the other Pierides by having only one very short vein omitted close to the apex of the wing from the third branch of the postcostal nerve; this peculiarity, not hitherto noticed by Lepidopterists, will further distinguish this genus from the black-veined white, in which the same typical arrangement exists, but the short nerve above mentioned is considerably longer and more distinct

From Gonipteryx Rhamni, in which the arrangement of the veins is nearly as in P Crataegi, they are at once separated by the form and colour of the wings From the orange-tipped butterfly they are distinguished by the shape and variegated colours of the wings, and especially by the apical veins of the fore wings, which are more numerous in that insect than in Pieris, the palpi also differ as well as the transformations The Bath white, which is united with the orange-tipped butterfly in the genus Mancipium, by Stephens and Curtis, is certainly referable to this genus, with which it is united by Boisduval and Ochsenheimer

The caterpillars are cylindric and fleshy, with numerous minute points, or larger tubercles, which emit pale hairs, and are arranged in regular transverse series The chrysalis is angulated with a short process in front of the head, and with a projecting lateral appendage behind (not in front of, as described by Stephens) each of the wing-cases They are generally to be found attached to walls by a little tuft of silk at the tail and by a girth round the middle of the body They do not constantly place themselves in one position with the head upright, but undergo this state in various positions

The genus is very extensive, the species being distributed over most parts of the globe, but especially in the intertropical parts of the old world, the Western hemisphere being comparatively poor in species In the great number of the species there exists a considerable number of natural divisions as pointed out by Boisduval and Lacordaire The caterpillars of such species as have been observed in the preparatory states, feed on the Cruciferæ, especially the species of Brassica, as well as on the Resedaceæ, Tropeolums, and Capparideæ They sometimes abound to a very great extent, especially when their natural enemies have failed, at such times our cabbages cauliflowers, &c become a prey to them, but their taste is so accommodating that they freely devour imported plants belonging to allied natural families The prevailing colour is white of a more or less clear hue, with a black edge at the extremity of the fore wings The females in our indigenous species are mostly marked with black spots in the posterior part of the disk of the fore wings Some of the exotic species are much more varied in their colours

The number of British species of this genus has been the subject of much recent inquiry, it having, until within the last fifteen years, been considered that there were but three—P Brassicæ, Rapi, and Napæ, exclusive of P Daplidice and Cardamines In 1827, however, Mr Stephens increased the number to seven in his Illustrations, separating P Chariclea from P Brassicæ, P Metra from P Rapæ, and P Napææ, and Sabellicæ from P Napi It is proper, however, to state that all these supposed new species had been indicated by previous authors, as will be noticed from the synonymes given below The propriety of their separation has, however, been questioned by several writers, amongst whom the Rev W T Bree has published some observations in opposition to the views of Mr Stephens in Loudon's Magazine of Natural History, vol 3, contending that as the three old established species are exceedingly variable, these supposed new species ought only to be regarded as varieties of them It is with the view of stating the grounds upon which these species rest, and of directing attention to the elucidation of the question that we have given figures of all the species, in some instances from the collection of Mr Stephens himself

No generic names have been so completely unsettled as the synonymous ones of Pontia and Pieris, a defect which has resulted from the want of some settled principle regulating the adoption of names of genera when an old established genus has been cut up into several others In the works of Stephens Curtis, and others, Pieris is the generic name used for the black-veined white, and Pontia for the garden white butterflies Boisduval, adopting the French mode of nomenclature, has employed the name Pieris for the genus of the garden whites, with which he has also associated the black-veined white and the Bath white butterflies, whilst he has given the name of Pontia to a few exotic species Ochsenheimer, on the other hand, in pursuance of the German nomenclature, has given all the whites, including the black-veined, the Bath white, the orange tip, and the small wood white, under the name of Pontia Such confusion as this is disgraceful, and ought to be no longer tolerated, although numberless other instances (in which the very commonest insects are known by three different generic names in England, France, and Germany) might be adduced to prove the necessity for some decisive step to remedy the evil In one respect naturalists (with the exception of a few vain persons who hesitate not to displace old specific names to make way for others of their own) are agreed in adopting the name first proposed for any genus or species, and to reject all subsequent ones proposed for the same group or species as synonymes In the present instance, therefore, by the adoption of this principle alone we shall be able to remedy the defect pointed out The name Pieris was first proposed by Schrank for all those butterflies which Linnæus had united together under the sectional name of Danai Candidi, including the brimstone and clouded-yellow butterflies with the whites, as well as the Parnassii About the same time Fabricius prepared the manuscript of his Systema Lepidopterorum in which he gave the name of Pontia to the black-veined white and garden whites, and that of Colias to the brimstone and clouded yellows But this work has never been published, and a very short generic abstract alone was given by Illiger some years afterwards in his Entomological Magazine Of these two last-mentioned names Colias is adopted by all entomologists for the clouded-yellow butterflies, and were we therefore to adopt the other Fabrician name Pontia, we should do injustice to Schrank, because Colias and Pontia are together synonymous with Pieris, which would thus be thrown out of use, but which on the contrary ought to be used for the great bulk of the white butterflies, other generic names being given to such aberrant species (but few in number) as may be required to be separated from the rest

In any case, the mode in which these names have been used has been erroneous for if we consider the black-veined-white as the type of Pieris, it is also the strict type of Pontia and therefore is synonymous therewith * or if on the other hand we regard the garden whites as the types of Pieris, as we contend ought to be done, the black-veined white ought either to receive a new generic name or be generically termed Pontia

In this view of the subject, and still further in order to remedy the confusion above alluded to, we consider it necessary entirely to reject the name Pontia except as a synonyme, to employ Pieris for the garden whites, as the natural types of the great body of the genus, and to give other generic names to such of the species of whites as have been separated therefrom

* Dalman, in 1816, united all the Danai Candidi into one genus named Ganoris

DESCRIPTION OF PLATE IV

INSECTS.—Fig 1 Pieris Brassicæ (the large garden white Butterfly), male 2 The female 3 The male showing the under side 4 The Caterpillar 5 The Chrysalis

" Fig 6 Pieris Chariclea (the early large garden white B) male 7 The female, showing the under side 8 The Caterpillar 9 The Chrysalis

PLANTS.—Fig 10 Brassica oleracea (Sea Cabbage) 11 Tropæolum majus 12 Tropæolum peregrinum

I have in this plate given three separate portraits of our commonest butterfly Pieris Brassicæ, in order to enable young collectors to ascertain at once the different markings and characters that distinguish this species from the other common white butterflies of the garden, which he would be apt to consider not worthy of attention, unless pointed out as distinct and well-defined species

On the same plate I have placed the insect most nearly resembling Pieris Brassicæ, viz Pieris Chariclea, which, however nearly resembling the former species possesses distinctive marks sufficient in the opinion of most entomologists to make it a separate species In addition to the difference of size, it will be seen that the black mark at the tips of the anterior wings is not indented with pointed arches as in P Brassicæ, and on the underside the secondary wings are not thinly sprinkled with black specks as in P Brassicæ, but thickly powdered with them, giving them a much more dusky appearance

No difference has as yet been satisfactorily ascertained in the respective caterpillars, which would determine the point whether P Chariclea is to be considered a distinct species or a mere variety But in Albin's old and curious work, P Chariclea is figured as the common cabbage and as its caterpillar and chrysalis are likewise figured, I have given accurate copies of them in the present plate If it could be ascertained that the caterpillar is the one from which the butterfly was raised it would assist in determining the question, as it presents some points of difference with that of P Brassicæ, as will be seen on comparison, and that it is the identical caterpillar which produced the butterfly figured with it is highly probable, as it appears from Albin's descriptions of his plates that it was his custom to make a drawing of a caterpillar immediately after taking it, another after it assumed the chrysalis form, and another on its perfect development, and this is the way in which his plates were produced, for he seems to have been so poor an entomologist, that even in the case of the common cabbage white butterfly, he would not have known what larva belonged to it by any other means Another reason for considering Albin's butterflies as produced from the caterpillars drawn is the circumstance that they made their appearance in the middle of April, thus agreeing with the time of appearance of P Chariclea

The plants in this plate are Brassica oleracea, the little plant from which all our garden cabbages cauliflowers, brocoli, &c &c , have been obtained by the arts of horticulture , Tropæolum majus erroneously known as the common *Nasturtium**, and Tropæolum peregrinum

I intend to confine myself to indigenous plants but in the present instance, the voracious appetites of the caterpillars represented, which greedily devour almost all succulent plants, have tempted me to introduce the Tropæolum majus to enliven the plate, and I was then induced to add the T peregrinum, as much less known, and as a hardy annual well worthy of general cultivation in our gardens H N H

SPECIES 1.—PIERIS BRASSICÆ THE LARGE GARDEN WHITE BUTTERFLY

Plate iv fig 1—5

SYNONYMES.—*Papilio (Danai Candidi) Brassicæ*, Linn Syst Nat ii 759 Donovan Brit Ins vol 13, pl 446 Lewin Brit Butt pl 25 Haworth	*Pieris Brassicæ*, Schrank, Latreille, Boisduval, Zetterstedt
Pontia Brassicæ, Fabricius, Ochsenheimer Leach, Stephens Curtis, Jermyn, Duncan Brit Butt pl 7, fig 1, 2	*Ganoris Brassicæ*, Dalman
	Catophaga Brassicæ, Hubner

This is one of the commonest species of butterflies, occurring in all parts of the country It varies in the expanse of its wings from two and a half to two and three-quarter inches The upper side of the wings is white the fore wings having a broad black patch occupying the apex of the upper side, being larger in the females than in the males, with its inner edge more or less distinctly notched , there is also a small black patch on the fore edge of the hind wings on the upper side The under side of the fore wings is marked in both sexes with two black

* When the Tropæolum was first introduced from Chili it was called Indian cress, from the circumstance of its being eaten in salads by the natives, and the name of *Nasturtium*, that of our common *water-cress* was inadvertently given to it

discoidal spots beyond the middle of the wings (which also appear on the upper side of these wings in the female), and the under side of the hind wings is dull yellow, covered with very minute black irroration The black apex of the fore wings is represented on the under side by a yellowish mark Sometimes, but rarely, the males have a black spot on the disc of the anterior wings above

The caterpillar of this species has been described as greenish yellow, with three yellow lines, but the entire ground colour of the caterpillar is uniformly greenish yellow, the segments being almost covered with black tubercles, varying in size and emitting white hairs (three of the larger ones on each side of each segment forming a triangle), but these tubercles are so placed as to leave a clear line on each side above the legs and one down the back, the head, fore legs, and anal segment, are also black The chrysalis is pale greenish spotted with black with three yellow lines

"The larvæ of this insect," observes Mr Haworth, "multiply so much in dry seasons as to make great havoc amongst our cabbages, &c Small birds destroy incredible numbers of them as food, and should be encouraged I once observed a Titmouse (Parus major) take five or six large ones to its nest in a very few minutes In inclosed gardens, sea-gulls with their wings cut are of infinite service I had one eight years, which was at last killed by accident, that lived entirely all the while upon the insects, slugs, and worms, which he found in the garden Poultry of any sort will soon clear a piece of ground, but unless they are of the web footed kind they do much damage by scratching the earth"—(Lepid Britann p 8) Great numbers of these caterpillars are also destroyed by a very minute species of Ichneumon-fly (Microgaster glomeratus), which deposits a considerable number of its eggs in the body of the living caterpillar, which are soon hatched, and produce minute footless grubs, that continue feeding upon the fat internal parts of the caterpillar, which, notwithstanding their presence, feeds on as though unconscious of any injury When, however, the time for its assuming the pupa state arrives it creeps up the adjacent walls, but instead of changing to a chrysalis, the little parasite grubs burst through its skin, which shrivels up into a small compass, arrange themselves close together by the side of the exuviæ of the caterpillar they have destroyed, and each spins for itself a little oval cocoon of yellow silk, which ignorant persons mistake for the eggs of the caterpillar and destroy, thus foolishly killing their benefactors In a short time the little parasites appear in their new form of active four-winged ichneumon flies

The transformations of this species have been carefully investigated by Swammerdam and Reaumur, whose researches, in conjunction with the anatomical details of the same species published by Herold, have left nothing to be desired on the subject The first of these authors chose this species to illustrate his history of "An Animal in an Animal, or the Butterfly hidden in the Caterpillar, which is a third particular example, serving as an additional illustration to the second method of the third order or class of natural transformations" The observations of this most indefatigable and celebrated author had for their object the proof of the natural production of insects from eggs laid by parents of the same species, and of the natural transformations of insects, and it is impossible to conceive more conclusive proofs, than are contained in his writings and figures, against the old theories of spontaneous generation and absolute metamorphosis

The eggs of this butterfly observed by Swammerdam are oval, with fifteen small longitudinal ridges converging to the centre of the smaller extremity of the eggs, the ribs themselves, and the membrane of the egg between them, being also divided crosswise by regular grooves or channels These eggs are deposited in clusters on the leaves of cabbages, &c, the larger end being applied to the surface of the leaf After several changes of the skin, the caterpillar prepares to undergo its change to the chrysalis state, and spins a little hillock of silk, which it seizes firmly with the hooks of its anal feet It has still, however, to construct a silken girth across the middle of its body, which it effects in a manner the most simple and least liable to accidents of the three modes adopted for this purpose by the different kinds of caterpillars which fasten themselves by girths The swallow tail butterfly presents us with one of these modes, in which, owing to the comparatively slight flexibility of the body, the caterpillar is forced to hold the skein of silk open by means of its fore legs The species of hair-streaked butterflies (Thecla) offer another mode, as will be detailed in our observations on that genus, but the caterpillars of this genus have a very flexible body, so that they are able to throw back the head until it extends to the back of the fifth segment of the body, its fore legs being elevated in the air, it then applies the spinneret of its lower lip to the surface on which it is stationed, close to one of the first pair of fleshy prolegs, and has only to carry its head over the body to the opposite side to fix the other end of the thread It then causes its head to return by the same route, emitting a second silken thread in like manner, one end of which it fastens at the spot at which the first was terminated, and the other end where the first was commenced By repeating this manœuvre a certain number of times, the skein of silk becomes sufficiently strong to bear the insect, and Reaumur states that it is composed of about fifty threads, as he had observed a caterpillar spin thirty-eight, and about a dozen had been already spun when he commenced the observation The number of these threads being completed, it only remains for the caterpillar to disengage its head from beneath the skein, a thing which might appear difficult, but which is easily effected by the caterpillar to effect this it brings the head close to the surface on one side where the threads are all fastened together, where in fact there is less liability to separate them from each other, which would be the case were the head to be withdrawn whilst it lies upon the middle of the back of the caterpillar, when the threads are of course loosest It is then carefully withdrawn The skein might be supposed to be too loose for the chrysalis, being spun over the body when that is doubled, but the future movements both of the caterpillar and chrysalis require that the skin should not be too tight, but should allow a little play in all directions, moreover the body, of course with the head turned back, as in the operation of spinning the skein, was stretched out, so that its natural diameter was considerably reduced Having thus completed its skein, it reposes quietly at full length, or rather its body contracts in length and becomes thicker, and at length the skin of the fore part of the back bursts and the head of the chrysalis appears, by continued writhing of the body the slit is enlarged and the skin pushed backwards beneath the skein of silk and thrown off at the tail, in the manner described under P Rapæ

This butterfly appears in the perfect state about the middle of May, or earlier if the weather be favourable according to Stephens It deposits its eggs at the end of the month, the caterpillars from which are soon hatched and feed together until the end of June, when they change to chrysalides, which period lasts from seven to about sixteen days (according to the heat of the weather) The perfect butterfly appearing therefore in July and depositing eggs which produce caterpillars which become full-fed so as to undergo their change to chrysalis in the autumn, in which state they remain till the following May It is very common throughout Europe, and is also found in Egypt, Barbary, Siberia, Nepaul, and even Japan The individuals from the two latter localities are, however, doubtfully regarded by Boisduval as distinct

SPECIES 2.—PIERIS CHARICLEA. THE EARLY LARGE WHITE BUTTERFLY.

Plate iv fig 6—9

SYNONYMES.—*Pontia Brassicæ Præcox*, Haworth MSS quoted in Steph Catal. | *Pontia Chariclea*, Steph Illust Brit Ent Illust 117 pl 3*, fig 2 Duncan Brit Butt pl 8, fig 1. *The Great White Butterfly*, Albin Ins pl 1.

This supposed species, "hitherto confounded with the preceding or unnoticed by entomologists, is considerably smaller than it," varying in expanse from two and a quarter to two and a half inches, and although with the same general style of marks, is distinguished from P Brassicæ by the dissimilar colour of the apical spot on the anterior wings above, which is ash coloured, without any internal indentations, and in the female deeply clouded with black, the cilia of all the wings is pale yellowish-white, the ground colour of the under side of the hind wings is more intense, and they are more deeply irrorated with black.

The time of the appearance of this insect is stated by Mr Stephens to be early in April, occurring in the same fields as P Brassicæ, and in an extensive series of large whites forwarded to him by the Rev W T Bree, captured between the 28th April and 23rd May, all the specimens taken previously to the 17th May agreed with Chariclea, whilst those subsequently taken were P Brassicæ, with the exception of a wasted P. Chariclea.

We agree with the arguments used by Mr Stephens against the opinion that the early appearance of these butterflies can be supposed to affect either their size, colour, or form of their markings, it still, however, remains to be proved whether these characters be constant or merely variable. We trust that our figures will be of service in directing attention to the specific rank of the butterfly.

DESCRIPTION OF PLATE V.

INSECTS.—Fig 1 Pieris Rapæ (the small garden white Butterfly), male 2 The female, showing the under side 3 The Caterpillar 4 The Chrysalis

Fig 5 Pieris Metra (Howard's white B), female 6 The male, showing the under side

" Fig 7 Pieris Napi (the green veined white B), male 8 The female, showing the under side 9 The Caterpillar 10 The Chrysalis

Fig 11 Pieris Sabellicæ (the dusky-veined white B), female 12 The male, showing the under side

PLANTS.—Fig 13 Brassica Rapa (Rape) 14 Brassica campestris (common navew)

Pontia Rapa and Pontia Napi, figured in this place, are both so obviously distinct from the species given in the previous plate, as well by their markings and character as from the decided difference of the caterpillars, that further remark is unnecessary in the description of the plate. Beneath P Rapæ I have placed the variety or sub-species known as P Metra, and under P Napi a corresponding variety or sub species distinguished by some as P Sabellica, the caterpillars of which have not been yet proved to differ. H N H

SPECIES 3.—PIERIS RAPÆ. THE SMALL GARDEN WHITE BUTTERFLY.

Plate v fig 1—4

SYNONYMES.—*Papilio (Dan Cand) Rapæ*, Linn Syst Nat ii 759 Haworth Lewin Brit Pap pl 26 Wilkes, pl 97. *Pontia Rapæ* Ochsenheimer, Stephens, Curtis, Duncan Brit Butt pl 7, fig 3. | *Pieris Rapæ*, Latreille, Boisduval, Zetterstedt. *Ganoris Rapa*, Dalman. *Catophaga Rapæ*, Hubner.

By persons ignorant of the nature of the growth and transformations of insects, this butterfly is considered as the young of P Brassicæ, with which indeed it exhibits considerable resemblance, although it is usually considerably smaller, varying, however, from one and two-thirds to nearly two and a half inches in expanse. It

Complete in One Volume 4to, price 2l. 2s. cloth; or 2l. 10s. half-bound morocco, gilt edges.

THE

LADIES' FLOWER-GARDEN

OF

ORNAMENTAL ANNUALS.

BY MRS. LOUDON.

ILLUSTRATED WITH FORTY-EIGHT CAREFULLY COLOURED PLATES.

Containing upwards of 300 Figures of the most showy and interesting Annual Flowers.

THE IMPERIAL CLASSICS.

On the 1st of September was published, Part II., price Two Shillings, of

BISHOP BURNET'S

HISTORY OF THE REFORMATION,

WITH HISTORICAL AND BIOGRAPHICAL NOTES.

To be completed in Thirteen or Fourteen Parts.

Just published, No. VI., price 2s. 6d.,

THE

LADIES' FLOWER-GARDEN

OF

ORNAMENTAL BULBOUS PLANTS.

BY MRS. LOUDON.

Each Number contains Three Plates, demy 4to size, comprising from Twelve to Twenty Figures accurately coloured from nature.

The whole will occupy about Twenty Numbers.

Price 3s. in cloth,

INSTRUCTIONS FOR COLLECTING, REARING, AND PRESERVING

BRITISH AND FOREIGN INSECTS.

AND FOR

COLLECTING AND PRESERVING SHELLS AND CRUSTACEA.

BY ABEL INGPEN, A.L.S. AND M.E.S.

A NEW EDITION, WITH THREE COLOURED PLATES.

(*To be completed in about Sixteen Numbers.*)

No. III.] [PRICE 2*s.* 6*d.*

BRITISH INSECTS AND THEIR TRANSFORMATIONS.

BRITISH BUTTERFLIES

AND

Their Transformations.

ARRANGED AND ILLUSTRATED IN A SERIES OF PLATES

BY H. N. HUMPHREYS, ESQ.

WITH CHARACTERS AND DESCRIPTIONS

BY J. O. WESTWOOD, ESQ., F.L.S.,

SEC. OF THE ENTOMOLOGICAL SOCIETY, ETC. ETC.

LONDON:

WILLIAM SMITH, 113, FLEET STREET.

MDCCCXL.

BRADBURY AND EVANS, PRINTERS, WHITEFRIARS.

NOTICE

The descriptions of Plate 7, 8, and 9, will be given in Number 4. As far as possible, care will be taken that the Engravings shall be placed in the same Number with their respective Descriptions, but as some species require a more lengthened notice than others, it will sometimes happen that this cannot be accomplished, as in the present instance

is of a creamy-white on the upper side, the tip of the fore wings having a very slight fuscous dusky or black irregularly defined mark, not extending along the entire margin of the wing On the under side this mark is replaced by a pale yellow mark, and the hind wings on the under side are also yellow, thickly spotted towards the base with black atoms The males have moreover a black spot, and the females two round spots on the upper side of the fore wings, and both sexes have two black spots on their under side The females have often also an elongated patch on the inner margin of the fore wings above, and there is a slight black mark on the costa of the hind wings All the markings vary greatly, and some females have the upper side dirty pale buff

If the perfect insects of this and the first species, P Brassicæ, thus agree so closely together, their preparatory states are totally unlike, affording reason to believe that the specific distinctions of the other presumed species might also be better determined by their preparatory states The eggs of this species are placed singly, and not in clusters, upon various species of Brassica, Reseda, &c , the caterpillars are pale green with a slender yellow dorsal line, and an interrupted yellow line above the feet on each side, in which the spiracles are placed, the head, feet, and tail, are also entirely green, the body is transversely wrinkled, and the segments are but slightly indicated Under a lens the whole body is seen to have a vast number of very minute black tubercles arranged in transverse rows It feeds on the interior leaves of the hearts of cabbages, &c , it is therefore much more obnoxious than P Brassicæ It is from this circumstance well known in France under the name of the *ver du cœur*

The mode in which the transformation of this insect from the caterpillar to the chrysalis state is effected, has been carefully described by Reaumur Having attached itself by the hind feet to the little bundle of silk at the tail, and suspended itself by a silken skein across the middle of the body, in the manner described in our observations on P Brassicæ it remains for about thirty hours unchanged, or rather the change is going on beneath the skin of the caterpillar, which gradually becomes of a dusky colour, owing to the separation of the body of the chrysalis within, the throwing off of the skin is effected very rapidly, "*c'est l'affaire d'un instant*" says Reaumur After a variety of contortions the skin of the back slits near the head, and forms a passage sufficiently large for the passage of the whole body of the chrysalis When the head of the chrysalis is disengaged it rests upon the old skin of the caterpillar, but it remains now to draw the hind part of the chrysalis out of the slit, or what is the same thing, to push back the caterpillar skin until it becomes a crumpled mass near the spot where the two hind legs are fixed, this is effected by the alternate lengthening and shortening of the chrysalis The skein of silk across the body is now seen to offer but little obstacle to the pushing back of the exuvia by the contraction of the rings of the chrysalis When the exuvia has been pushed back so as to cover only about one-third of the length of the chrysalis, the insect ceases this operation, it being more convenient for it to withdraw the extremity of the chrysalis out of the aperture (being upheld by the transverse girth) and then push it back outside the exuviæ till it reaches the little bundle of silk, into which it fixes the little hooks at the extremity of the tail It then rids itself of the exuviæ by a semicircular movement of the extremity of the abdomen, which is at the time curved, whereby it pushes the exuvia out of its old position and breaks the threads of silk to which it was attached, when it falls down The chrysalis then remains quiet, being fixed exactly in the same situation and manner as the caterpillar The chrysalis is yellowish, greenish-grey, or brownish, often with three sulphur yellow lines

According to Mr Stephens, the first brood appears at the end of April and the second about the beginning of July, but there is evidently no regularity in the broods, as may be seen from the following result of the observations of Jacob L'Admiral of the dates in which specimens of this species spun their webs, became pupæ, and appeared as butterflies, with the number of days in which they remained as pupæ

SPON		WIERD EEN POPJE	EN KAPEL	IN DAGEN
De eerste,	18 July, 1720	20 July	5 Augustus,	— 15
De 2de	6 July, [Juny ?] 1720	8 Juny	19 Juny,	— 11
De 3de,	13 September	16 September	1 April,	— 197
De 4de,	2 September, 1739	4 September	23 September	— 19
De 5de,	2 September	5 September	28 May	— 265

Like P Brassicæ this species inhabits the whole of Europe, from Lapland to the south, and is found in Egypt, Barbary, Asia Minor, Siberia, and Cachmere The P Ergane, Hubn (P Narice of Dahl and Treitschke), and the P Nelo of Borkhausen, are probably varieties of this species

---

SPECIES 4—PIERIS METRA HOWARD'S WHITE BUTTERFLY

Plate v fig 5, 6

SYNONYMS—*Pontia Metra*, Stephens' Illust Brit Ins Haustell, vol i p 19 and 146 Duncan Brit Butt pl 3 fig 2

Male—*Papilio alba media immaculata*, Petiver Pap pl 1, fig 13, 14

Female—*Papilio alba media trimaculata*, Petiv pl 1, fig 11 12

This supposed species bears the same relation to P Rapæ as P Chariclea does to P Brassicæ, differing from it in its smaller size, varying from twenty to twenty five lines in expanse, and being also early in appearance, the first brood being found early in April, or even at the middle of March In the appendix to his work, Mr Stephens questions whether this or P Chariclea be double-brooded, as out of several hundred specimens taken in July and August, not one belonged either to P Metra or Chariclea but to P Rapæ and Brassicæ P Metra is very variable in its markings but is generally distinguished from P Rapæ by the comparative slenderness and truncation of the fore wings, which are consequently very acute at the apex, which is slightly clouded with dusky, and by the black base of the wings "The male has a single obsolete dusky spot, and the female two, that at the anal angle being geminated, this sex has also the basal half of the wing much clouded with dusky the posterior wings in both sexes are white, with the base black, and a dusky costal spot Beneath, the sexes are similar, the anterior wings are white, with the tip yellow, the base and two obsolete spots dusky, the posterior wings are bright-yellow with a pale-orange streak on the costa, strongly irrorated throughout with dusky, the anterior half of the discoidal cell being least speckled, the antennæ, legs, and body, resemble those of P Rapæ, the cilia are entirely clear, white" (Stephens) Mr Stephens describes two varieties of the male, in which the markings are less distinct, or even almost entirely obliterated He also in his appendix mentions several circumstances in support of the specific distinction of these early "small whites," and states that the present had long been known amongst collectors under the name of "Mr Howard's white" If indeed it be admitted that these 'early whites' exhibit such distinctions of form, size, colour, and markings, it seems impossible to suppose that their detention in the chrysalis through the winter months should have the effect of producing such striking peculiarities The caterpillar has not unfortunately been yet observed, 'but the chrysalis does not materially differ from that of P Rapæ" (Stephens)

SPECIES 5.—PIERIS NAPI. THE GREEN-VEINED WHITE BUTTERFLY.

Plate v fig 7—10

SYNONYMES.—*Papilio (Dan. Cand.) Napi*, Linn. Syst. Nat. ii. 760. Lewin Brit. Pap. pl. 27. Donovan Brit. Ins. vol. viii. pl. 280, fig. 1. Albin Ins. pl. 52, fig. d—g. Wilkes Ins. pl. 98.
Pontia Napi, Fabricius, Ochsenheimer, Stephens, Curtis, Duncan Brit. Butt. pl. 9, fig. 1.
Pieris Napi, Schrank, Latreille, Boisduval, Zetterstedt.
Ganoris Napi, Dalman.
Cataphaga Napi, Hubner.

This species is at once distinguished by the green colour of the parts of the wings adjoining the veins on the under side, whence the English name of the species. It varies in the expansion of its wings from one and one-third to two inches. The wings are white above, except at the base, which is generally black; the tip is also dusky; the apex of the branches of several of the subcostal nerves marked with small triangular spots. The males have generally a black spot between the middle and apex of the wing. The females have the apical dark mark larger, and have also two large black spots towards the posterior margin, the hinder one being connected with a black dash on the inner margin. On the under side the fore wings of the males have the tip yellow, the nervures dusky, and two black spots corresponding with those of the females. In both sexes the hind wings are pale-yellow beneath, with the veins broadly margined on each side with dusky-greenish, and the hind wings have a small dusky mark on the costal edge above. The fore wings in the females are more rounded than in the males.

This is a very common and very variable insect, being found especially at the middle of May and beginning of July, in gardens and pastures, the larva feeding on the Navew and other species of Brassica, Reseda, Raphanus, and other plants. It is pubescent, of an obscure green colour on the back, but brighter on the sides, with the spiracles red, placed upon a small yellow spot on each segment. The chrysalis is greyish or yellowish green, with black spots.

Amongst the varieties of this insect should also most probably be arranged the *Papilio Napææ* of Esper, distinguished by being of a larger size than the ordinary specimens of P. Napi, varying from $1\frac{1}{6}$ to $2\frac{1}{6}$ inches in expanse; "the male has the upper surface of the wings milk-white, with the tip, a spot, and two or three triangular dashes on the hind margin of the anterior, black; beneath, the latter have slightly dilated greenish nervures, with two cinereous spots placed transversely, and a yellowish tip; the posterior wings are pale-yellowish, with a deeper costal streak; the basal nervures above dilated and greenish. The female has the tip of the anterior wings and three spots, one of which is subtriangular, and placed on the thinner edge of the wings, black or dusky, and the posterior wings are clearer yellow. The nervures on the under surface of the posterior wings are more or less dilated in different specimens." Such is the description given by Mr. Stephens, who however adds, "I think with Godart that it may only be a very large variety of P. Napi, but as it appears to have characters sufficient to constitute a distinct species, the determination of this point must be left for future investigation." The caterpillar and chrysalis have not been observed, nor have any circumstances connected with the time or place of its appearance been given, so that we have less ground for considering it as distinct than exists in respect to P. Chariclea and Metra.

SPECIES 6.—PIERIS SABELLICÆ. THE DUSKY-VEINED WHITE BUTTERFLY.

Plate v. fig. 11, 12.

SYNONYMES.—*Papilio Sabellicæ*, Petiver Papil. pl. 1, fig. 17, 18, ♂ fig. 15, 16, ♀.
Pontia Sabellicæ, Steph. Illus. Haust. I. p. 22, pl. 3*, fig. 3, 4. Duncan Brit. Butt. pl. 8, fig. 3.
P. Bryoniæ, Godart Enc. Meth. 9, p. 162, No. 146.
P. Napi, var. Haworth, Boisduval, Zetterstedt.

This supposed species differs from P. Napi (to which it is otherwise closely allied in general markings and appearance), not only in having the veins on each side strongly margined with brown but in the form of the fore wings, which Mr. Stephens describes as being nearly of the form of those of P. Cardamines; but this is not quite correct, the apex of these wings being as acute as in P. Napi, but the lower portion of the outer margin of the wing is greatly dilated. In my specimen of the female, the difference in the form of the wings when compared with a male of P. Napi (in which the expansion, measured from tip to tip, is precisely similar) is most striking; for on measuring the expanse from the base to the apex of the middle branch of the medial wing-vein of this specimen of P. Sabellicæ, each wing is found to be more than 1½ line longer than the same portion of the wing in P. Napi, thus making a difference of nearly a quarter of an inch across this part of the wing, although the measure between the tips is alike. It is true that the females of P. Napi are described as having the "anterior wings more rounded than the male," but Mr. Stephens also describes and figures a male with the characters of P. Sabellicæ. Such a character, if found permanently in conjunction with the dark margins of the veins of the wings both on the upper and under surface, we should certainly deem of specific value, although Boisduval gives it as a variety of P. Napi, describing only the female; whilst Zetterstedt describes both sexes of this as variety B. of P. Napi, but adds that the females were most abundant. He however states, that he had repeatedly captured the true males of P. Napi united with the females of his variety B. (or P. Sabellicæ). He had also reared a female of his var. B. from a yellowish-green chrysalis, very similar to the ordinary chrysalis of P. Napi, on the 20th of June.

DESCRIPTION OF PLATE VI.

INSECTS.—Fig. 1. Euchloe Cardamines (the orange-tip Butterfly), male. 2. The female. 3. The female, showing the under side. 4. The Caterpillar. 5. The Chrysalis.

" Fig. 6. Pieris Daplidice (the Bath white B.), male. 7. The female. 8. The male, showing the under side. 9. The Caterpillar. 10. The Chrysalis.

" Fig. 11. Leptoria Candida (the wood-white B.). 12. The Caterpillar. 13. The Chrysalis.

PLANTS.—Fig. 14. Reseda lutea (base Rocket, or wild Mignonette). 15. Cardamine pratensis (common meadow Ladies Smock). 16. Lathyrus pratensis (meadow Vetchling).

As the male and female of two of the species represented in the above plate present marked differences, I have in both instances given portraits of both sexes, as well as an under side view which will enable the young collector to distinguish at once the female of E. Cardamines from the male of P. Daplidice, which from their similarity in some respects he might have otherwise mistaken. The figures of P. Daplidice are from Italian specimens in my own collection, which differ in no respect from the indigenous specimens which I have seen. The E. Cardamines and L. Sinapis are from British specimens, and I shall in future endeavour to have all the plates drawn from British specimens only. H. N. H.

SPECIES 7.—PIERIS DAPLIDICE. THE BATH WHITE BUTTERFLY.

Plate 1 fig 6—10

SYNONYMES.—*Papilio (Dan. Cand.) Daplidice*, Linn. Syst. Nat. ii. 760. Lewin Brit. Papil. pl. 28. Donov. Nat. Hist. Brit. Ins. 6. pl. 200.

Pontia Daplidice, Fabricius, Ochsenheimer, Steph. Ill., Curtis Brit. Entomol. pl. 48.

Mancipium Daplidice, Stephens Nom., Duncan Brit. Butt. pl. 9, fig. 2.

Pieris Daplidice, Schrank, Latreille, Boisduval, Zetterstedt.

Synchloe Daplidice, Hubner (Verz. bek. Schmet.).

The Slight greenish Half-mourner, Petiver Pap. pl. 2, fig. 8, male.

Vernon's Greenish Halfmourner, Petiver Pap. pl. 2. fig. 9 female.

This very rare butterfly varies in the expansion of its wings from an inch and two-thirds, to nearly two inches. The wings are of a white colour, those of the males being rather more cream-coloured. The upper side of the fore wings is blackish at the base, and is marked with a rather large, discoidal black spot at the extremity of the discoidal cell, in which the transverse veins appear of a white colour. The apex of these wings is irregularly black, the dark colour being broadest towards the front margin, extending only to the middle branch of the mediastinal vein, and being irrorated with white, having also four irregular white spots in the black patch, which is darker in the females than in the males, the females have moreover a small black patch near the inner margin of the fore wings; the upper side of the hind wings is white but exhibits traces of the marking on the under side, in consequence of their slight transparency; indeed in the female, these traces are more or less distinctly marked with black scales, especially along the edge. On the under side the markings are alike in both sexes, the male having the spot on the inner edge of the fore wings, which is wanting on the upper side in this sex. The marks of the fore wings on the under side are of a greenish colour. The under side of the hind wings is yellowish-green or greenish (in some females), with three large white spots forming a triangle towards the outer base of the wing, succeeded by an irregular white bar beyond the middle of the wing, traversed by yellowish veins and with five white clavate spots on the outer margin. The male moreover differs from the female in the form of the fore wings, which are more acute at the apex than in any other species of this genus * and with the external margin slightly concave, instead of being convex, as in the female. This remarkable sexual difference, hitherto I believe unnoticed in this species, occurs as we have seen, but in a less striking degree, in P. Napi and Sabellicæ. The caterpillar, which feeds upon various wild Resedacea and Cruciferæ, such as wildwoad, base rocket, and cabbage, is, according to Boisduval, of an ashy blue colour, covered with small black granules, with four white longitudinal stripes, marked at each incision with a lemon-coloured spot. The belly and legs are whitish, with a yellow spot above each of them. The chrysalis is greyish dotted with black, with several reddish stripes. Our figures of the larva and pupa (carefully copied from Hubner) differ in their colours from the individuals described and figured by Boisduval in his 'Collection Iconographique des Chenilles d'Europe.'

On the Continent, two broods of this insect appear in the course of the year †. It is found in dry and sandy situations, and is very common, especially in the more southern part of the continent of Europe, as well as in Barbary, Asia Minor, and Cachmere. In this country it is very rare. According to Ray, it was formerly taken by Vernon near Cambridge, and Petiver records it as having been taken near Hampstead. Lewin informs us

---

* Notwithstanding this circumstance, the species is placed by Stephens at the head of his section division of Pontia characterised by having "the anterior wings distinctly rounded at the tip," in contradistinction from the typical species, P. Brassicæ &c., in which they are described as "obtusely angled."

† In April and May and afterwards in August, according to Godart, but Boisduval gives April and May and June and July, as the time of its appearance. Mr. Stephens captured his specimen however in the middle of August.

that it was named the Bath White, from a piece of needle-work executed at Bath by a young lady, from a specimen of this insect said to have been taken near that city Mr Haworth, in his Lepidoptera Britannica, (pref p xxvi) states, that it had been taken in the preceding *June* (not May as mentioned by Stephens) in White Wood, near Gamlingay, Cambridgeshire Mr Stephens captured it on the 14th August 1818, in the meadow behind Dover Castle, where other specimens have since been captured (Entomol Mag iii. 400) According to Mr Dale, a specimen was also taken about the same time near Bristol

Our recent English entomologists have been singularly unfortunate in respect to the generic relations of this insect Mr Curtis first proposed forming it and P Cardamines L, into a section of the genus Pontia, having the wings variegated beneath, and the terminal point of the palpi shorter than the second Subsequently, Mr Stephens, although noticing Mr Kirby's observations on the peculiarity of the metamorphosis of P Cardamines, adopted this section under the name Mancipium—Hubner In none of his characters of the section, however, except the trivial one of the variegated under surface of the wings, is there any agreement between P Daplidice and Cardamines The labial palpi in a female of Daplidice which I have dissected, have the second and third joints of equal length, although Savigny and Curtis figure the third as scarcely more than half the length of the second, the anterior wings of the female Daplidice are not rounder than those of P Napi, whilst we have seen that they are much more acute in the males At the same time, the character of the transformations of the two species is totally distinct (see Plate VI those of P Daplidice agreeing with the rest of the genus) Moreover, the antennæ in Daplidice are terminated by a suddenly formed flat club, which is even broader than in P Napi, whilst in P Cardamines the club is long, and gradually formed, and lastly, the veins of the fore wings * are arranged as in Pieris Napi, differing from those of P Cardamines

GENUS VI
EUCHLOE†, Hubner. ANTHOCHARIS, Boisduval.

This genus is closely allied to Pieris in many respects. but differs in others, which are considered as of primary importance, the palpi are especially distinct, the second joint being very long and the third very minute, not being more than one-fourth of the length of the preceding ‡ The antennæ are short, and terminated by a gradually formed oval compressed club, the fore-wings are much more dilated and rounded at the tips, and in order to support this increased expanse, the third branch of the postcostal vein emits from its upper side two distinct veins § The under wings form a slight channel for the abdomen The wings are somewhat transparent so as to show the markings of the under side when viewed from above They are generally ornamented with a bright orange spot at the tip in one or both sexes, and on one or both of their surfaces

* In the specimens of P Daplidice, the wings of which I have examined and denuded of scales, the third branch of the postcostal nerve is destitute of the short branch *emitted* (not *omitted* as misprinted in p 19, line 6) close to the apex of the fore wing in the other species of Pieris, but the position of this little branch varies in the different species of Pieris, and it is sometimes even wanting, as I have more recently discovered, indeed, I possess specimens, one fore wing of which possesses this short branch, and the other wants it Typically speaking however it is the character of Pieris

† Derived from the Greek εὐ very, and χλοη the green herb in allusion to the spotted green wings

‡ Boisduval describes the last joint of the palpi as "a peine aussi long que le précédent "—(Hist Nat Lep 1, 556) But my description is taken from a carefully dissected specimen of E Cardamines

§ Boisduval figures Anthocharis Autevippe (Hist Nat Lep pl 18, fig 3) with only one vein from this third branch, as in Pieris

The caterpillars are much more slender than those of Pieris, pubescent and attenuated at each end of the body. The chrysalis is naked, strongly boat-shaped, and more or less curved, pointed at each end, destitute of lateral points. Boisduval states that the segments of the chrysalis are unmoveable.

SPECIES 1.—EUCHLOE CARDAMINES. THE ORANGE-TIPPED BUTTERFLY.

Plate vi fig 1—5

SYNONYMES.—*Papilio (Dan. Cand.) Cardamines*, Linn. Syst. Nat. ii. pl. 761. Lewin's Papil. pl. 30. Harris's Aurelian pl. 22, fig. g. h. Wilkes Brit. M. & B. pl. 99. Donovan Brit. Hist. pl. 169. *Mancipium Cardamines*, Stephens. Duncan, Brit. Butt. pl. 10, fig. 1, 2.

Anthocharis Cardamines, Boisduval, Godart. *Pieris Cardamines*, Schrank, Latreille and Zetterstedt. *Ganoris Cardamines*, Dalman. *Euchloe Cardamines*, Hubner. Verz. bek. Schm. 1816.

This beautiful insect varies in the expanse of its wings from $1\frac{1}{3}$ to 2 inches. The ground colour of the wings is white; on the upper side the base is black; there is a black semicircular mark at the extremity of the discoidal cell, and the apex is black, with pale spots along the margin. In the males, the space between the discoidal spot and the dark apex is suffused on both surfaces with bright orange, which is entirely wanting in the females; the upper surface of the hind wings is white, but exhibits the traces of the markings which ornament the under side; the cilia is marked with seven black dots. The fore wings on the under side are yellowish at the base, and the dark apical spot is pale grey varied with green: the under wings on this side in both sexes are marbled with white and green, the veins being edged with yellow.

Mr. Stephens describes several varieties,—one having a black spot on the upper surface of the hind wings, and another with the black lunule of the fore wings almost obliterated. Mr. Haworth also describes a variety of the male having the orange spot almost obliterated above; and Boisduval mentions a variety of the female having an orange spot on the under side of the fore wings.

The larva is green, slightly pubescent, very finely dotted with black, with a white lateral stripe. It feeds upon Cardamine impatiens (whence its specific name), Turritis glabra, Brassica campestris, &c. The chrysalis is at first green, but in a few days it assumes a yellowish grey colour with brighter coloured stripes.

This is a very abundant species, sporting about sunny lanes and pastures, and open places in woods, in the early spring. Mr. Stephens states that of six pupæ of this species, two came to perfection at the end of May, one in the beginning and one at the end of June, and the other towards the middle of July: thus accounting for the apparently long continuance of the insect in the perfect state.

GENUS VII.

LEPTORIA,* HUBNER. LEUCOPHASIA, STEPHENS.

This genus is at once distinguished from its allies by the very peculiar form of the wings, which are narrow, elongated, and slender, the anterior being rounded at the tip, and the posterior slightly grooved. The head is of moderate size, with the eyes large and prominent; the palpi short, the basal joint being very large and broad, the second small and square, and the third very small and rather oval; the antennæ are terminated by a rather

* Evidently derived from the Greek λεπτος, slender, in allusion to the delicate form of the wings.

abruptly formed obconical compressed club, the thorax is very small, and the abdomen very long and slender. Like its allies, the wings of the female are more rounded at the tip than those of the male. The discoidal cell in both the fore and hind wings is very short, scarcely extending beyond one-fourth of the length of the wing from the base, this cell is closed by a transverse vein. The arrangement of the veins of the fore wings is peculiar, differing from every other English butterfly. The postcostal vein instead of emitting several distinct branches in front, emits but one which branches off at the apex of the discoidal cell, but this branch emits four veins in front so that in effect an equal support is given to the membrane of the wing, as though these veins had separately branched off from the main postcostal vein.* The ungues are distinct and bifid.

The caterpillar is slender, attenuated at each end, very slightly pubescent. The chrysalis is angulated, spindle-shaped, nearly resembling that of E. Cardamines, but not bent in the middle, and with the segments moveable.

The habits of the species materially differ from those of Pieris, the perfect butterfly frequenting woods, as its English name indicates, and its larva feeding on leguminous herbs.

SPECIES 1.—LEPTORIA CANDIDA, *Westwood*

Plate vi fig. 11—13

SYNONYMES.—*Papilio (Dan. Cand.) Sinapis* Linn. Syst. Nat. ii 760. Lewin Papil. pl. 29. Donovan Brit. Ins. vol. viii pl. 280, fig. 2. Harris Aurel. pl. 29, fig. t. u.
Pontia Sinapis, Fabr. Ochs., Leach.
Pieris Sinapis Schrank, Latreille, Godart.
Leucophasia Sinapis Stephens, Boisduval. Duncan Brit. Butt. pl. 10 fig. 3.
Candonis Sinapis Dalman, Zetterstedt.
Leptoria Sinapis, Hubner, Verz. bek. Schmett.
Leucophasia Loti Renner Conspect. p. 4.
Papilio Candidus Retzius (Gen. et Sp. Ins. De Geer) p. 30.

This delicate little butterfly varies in the expanse of its wings from $1\frac{1}{3}$ to $1\frac{5}{6}$ inches. The wings are of a pure white colour on the upper side, with a roundish dark blackish spot occupying the tip of the fore wings, and which in the females is of a paler greyish colour. In some specimens, however, this apical patch is entirely wanting.* The under side of the fore wings has the fore margin greyish coloured, interrupted by a more or less distinct whitish crescent-like mark, placed at the extremity of the discoidal cell, the base and tip of these wings being of a very pale yellowish green; the under side of the hind wings is slightly stained with greenish yellow, with the veins and two irregular and often interrupted transverse bars of a greyish ash colour.

The caterpillar is described by Boisduval as green, with the dorsal vessel rather darker, and a lateral yellow stripe situated above the feet. The chrysalis is at first of a greenish yellow, but subsequently of a whitish grey with red dots on the sides and upon the wing cases (Collect. Iconogr. des Chenilles d'Europe).

The caterpillar feeds upon Vicia cracca, and on the species of Lotus, Lathyrus and Orobus growing in woods. Linnæus, however, says of it "Habitat in Brassica et affinibus," and Fabricius, "Habitat in Brassica Sinapi Rapa," and hence the name of P. Sinapis was given to it. As, however, such is now ascertained not to be the case, the insect confining its attacks to a very different natural order of plants, I have thought it more proper to refer to the name proposed by Retzius, the commentator of De Geer, rather than to give it a new specific name.

Its flight is slow and undulating; it is by no means a common insect, frequenting the glades in woods of the southern counties of England, and appears to be double brooded, being found in the winged state in the middle of May and beginning of August.

---

* The same arrangement of the veins is found in Euterpe Charops, figured by Boisduval (Hist. Nat. Lep. pl. 18, fig. 2).

ILLUSTRATED EDITION OF FROISSART.

In two thick Volumes, price 36*s*

SIR JOHN FROISSART'S CHRONICLES OF ENGLAND, FRANCE, SPAIN, &c.

THIS Edition is printed from the Translation of the late THOMAS JOHNES, Esq, and collated throughout with that of LORD BERNERS, numerous additional Notes are given, and the whole embellished with *One Hundred and Twenty Engravings on Wood*, illustrating the Costume and Manners of the period, chiefly taken from the illuminated MS copies of the Author, in the British Museum, and elsewhere

---

COMPANION TO FROISSART.

In two Volumes, price 30*s*

THE CHRONICLES OF MONSTRELET.

WITH NOTES AND WOODCUTS, UNIFORM WITH THE ABOVE EDITION OF FROISSART

ILLUSTRATED EDITION OF "MARMION"

In demy 8vo, price 16*s* cloth, 21*s* morocco elegant,

MARMION.

A Poem

BY SIR WALTER SCOTT

WITH FIFTY BEAUTIFUL WOOD ENGRAVINGS

ILLUSTRATED WITH FIFTY-ONE PORTRAITS,

BURNET'S HISTORY OF HIS OWN TIMES.

In Two Volumes, super-royal 8vo, cloth lettered, price 2*l* 2*s*, or half bound in morocco, 2*l* 12*s* 6*d*

ILLUSTRATED WITH FIFTY-SIX PORTRAITS,

CLARENDON'S HISTORY OF THE REBELLION,

In Two Volumes, imperial 8vo, price 2*l* 10*s* cloth lettered

---

In one Volume, foolscap 8vo, price 7s in cloth,

A TREATISE ON THE INSECTS INJURIOUS TO THE GARDENER, FORESTER, AND FARMER.

TRANSLATED FROM THE GERMAN OF M KOLLAR, AND ILLUSTRATED WITH ENGRAVINGS

BY J AND M LOUDON.

WITH NOTES BY J O WESTWOOD, ESQ

"We heartily recommend this treatise to the attention of every one who possesses a garden, or other ground, as we are confident that no one taking an interest in rural affairs can read it without reaping both pleasure and profit from its perusal"—*Literary Gazette*

"We have always wondered that, in a country like this, where the pursuits of agriculture and horticulture are so universal and important, entomologists should never have bethought them of writing a book of this description It is therefore, with great satisfaction, that we announce the appearance of the present translation of a work which goes far to supply the deficiency we have spoken of"—*Athenæum*

"From the very neat and cheap manner in which the volume is got up, we trust it will become a favourite, not only with the entomologist, but with every lover of agriculture, arboriculture, and horticulture"—*Mag Nat Hist*, Feb 1840

No. IV.] [PRICE 2s. 6d.

BRITISH INSECTS AND THEIR TRANSFORMATIONS.

BRITISH BUTTERFLIES

AND

Their Transformations.

ARRANGED AND ILLUSTRATED IN A SERIES OF PLATES

BY H. N. HUMPHREYS, ESQ.

WITH CHARACTERS AND DESCRIPTIONS

BY J. O. WESTWOOD, ESQ., F.L.S.,

SEC. OF THE ENTOMOLOGICAL SOCIETY, ETC. ETC.

LONDON:
WILLIAM SMITH, 113, FLEET STREET.

MDCCCXL.

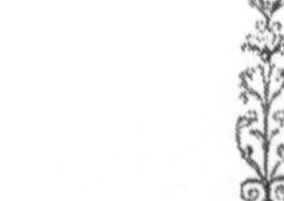

BRADBURY AND EVANS, PRINTERS, WHITEFRIARS.

WORKS PUBLISHED BY WILLIAM SMITH, 113, FLEET STREET.

SMITH'S STANDARD LIBRARY.

In medium 8vo, uniform with Byron's Works, &c

Works already Published

THE LADY OF THE LAKE 1s
THE LAY OF THE LAST MINSTREL. 1s
MARMION 1s 2d
THE VICAR OF WAKEFIELD 1s
THE BOROUGH BY THE REV GEORGE CRABBE 1s 4d
THE MUTINY OF THE BOUNTY 1s 4d
THE POETICAL WORKS OF H KIRKE WHITE 1s
THE POETICAL WORKS OF ROBERT BURNS 2s 6d
PAUL AND VIRGINIA, THE INDIAN COTTAGE, AND ELIZABETH. 1s 6d
MEMOIRS OF THE LIFE OF COLONEL HUTCHINSON (GOVERNOR OF NOTTINGHAM CASTLE DURING THE CIVIL WAR) BY HIS WIDOW MRS LUCY HUTCHINSON 2s. 6d
THOMSON'S SEASONS, AND CASTLE OF INDOLENCE. 1s
LOCKE ON THE REASONABLENESS OF CHRISTIANITY 1s
GOLDSMITH'S POEMS AND PLAYS 1s 6d
KNICKERBOCKER'S HISTORY OF NEW YORK 2s 3d
NATURE AND ART BY MRS INCHBALD 10d
SCHILLER'S TRAGEDIES THE PICCOLOMINI, AND THE DEATH OF WALLENSTEIN 1s 8d
ANSON'S VOYAGE ROUND THE WORLD. 2s 6d
THE POETICAL WORKS OF GRAY AND COLLINS. 10d
THE LIFE OF BENVENUTO CELLINI 3s.
HOME BY MISS SEDGWICK 9d
THE VISION OF DON RODERICK, AND BALLADS BY SIR WALTER SCOTT 1s
TRISTRAM SHANDY BY LAURENCE STERNE 3s
STEPHENS'S INCIDENTS OF TRAVEL IN EGYPT, ARABIA PETRÆA, AND THE HOLY LAND 2s 6d
STEPHENS'S INCIDENTS OF TRAVEL IN GREECE, TURKEY, RUSSIA, AND POLAND 2s 6d
BEATTIE'S POEMS AND BLAIR'S GRAVE 1s
THE LIFE AND ADVENTURES OF PETER WILKINS 2s. 6d
POPE'S HOMER'S ILIAD 3s
UNDINE A MINIATURE ROMANCE FROM THE GERMAN. 9d
ROBIN HOOD, REPRINTED FROM RITSON 2s 6d
LIVES OF DONNE, WOTTON, HOOKER, HERBERT, AND SANDERSON BY IZAAK WALTON 2s 6d
LIFE OF PETRARCH. BY MRS DOBSON 3s
GOLDSMITH'S CITIZEN OF THE WORLD 2s 6d
MILTON'S PARADISE LOST 1s 10d
MILTON'S PARADISE REGAINED AND MINOR POEMS 2s
RASSELAS BY DR JOHNSON. 9d
ALISON'S ESSAYS ON TASTE 2s 6d
A SIMPLE STORY BY MRS INCHBALD 2s
THE POETICAL WORKS OF JOHN KEATS. 2s
GOLDSMITH'S ESSAYS 2s

THREE VOLUMES ARE NOW ARRANGED ON THE FOLLOWING PLAN —

ONE VOLUME OF "POETRY,"

CONTAINING

SCOTT'S LAY OF THE LAST MINSTREL
SCOTT'S LADY OF THE LAKE
SCOTT'S MARMION
CRABBE'S BOROUGH
THOMSON'S POETICAL WORKS
KIRKE WHITE'S POETICAL WORKS
BURNS'S POETICAL WORKS

ONE VOLUME OF "FICTION,"

CONTAINING

NATURE AND ART BY MRS INCHBALD
HOME BY MISS SEDGWICK
KNICKERBOCKER'S HISTORY OF NEW YORK BY WASHINGTON IRVING
PAUL AND VIRGINIA BY ST PIERRE
THE INDIAN COTTAGE BY ST PIERRE
ELIZABETH BY MADAME COTTIN
THE VICAR OF WAKEFIELD BY OLIVER GOLDSMITH
TRISTRAM SHANDY BY LAURENCE STERNE

ONE VOLUME OF "VOYAGES AND TRAVELS,"

CONTAINING

ANSON'S VOYAGE ROUND THE WORLD
BLIGH'S MUTINY OF THE BOUNTY
STEPHENS'S TRAVELS IN EGYPT, ARABIA PETRÆA, AND THE HOLY LAND
STEPHENS'S TRAVELS IN GREECE, TURKEY, RUSSIA, AND POLAND

These Volumes may be had separately, very neatly bound in cloth, with the contents lettered on the back, price 10s 6d each

Those parties who have taken in the different works as they were published, and who wish to bind them according

DESCRIPTION OF PLATE VII

INSECTS.—Fig 1 Parnassius Apollo (the Apollo or crimson ringed Butterfly) 2 Showing the under side 3 The Caterpillar
4 The Chrysalis See pages 11 and 12 for the description of this insect
" Fig 5 Aporia Cratægi (the black veined white B) 6 The female 7 The Caterpillar 8 The Chrysalis
PLANTS.—Fig 10 Sedum Telephium 11 Cratægus oxyacantha (Hawthorn)

Though not in the order of the latest arrangements, I have given Aporia Cratægi and Parnassius Apollo together, as from many features of resemblance, the semitransparency of a portion of the wings, &c &c, they group better than they would have done in their respective places The remarkably perfect specimen from which Parnassius Apollo is drawn, I took upon the Mount Cenis a few summers ago in the month of July, where it is found in such abundance that I captured between thirty and forty specimens in less than half an hour Aporia Cratægi, male and female, are from specimens taken near Tivoli The under side of Aporia Cratægi is generally described as closely resembling the upper surface, but in my female specimen, in addition to the brown veinings which usually distinguish it from the male, the anterior wings are of a deeper cream-colour, and the posterior ones inclining to pale yellow The larva and pupa are both from Hubner Of the plants I have only to remark, that the red variety of the hawthorn is given instead of the more common white, as affording a better contrast to the colours of the insects Both the caterpillars are from Hubner H N H

GENUS VIII.

APORIA* HUBNER PIERIS, STEPHENS

This genus is closely allied to the garden white butterflies, but is distinguished from them not only by its habits subsequently detailed but also by various peculiarities in its structure The palpi are rather short, with the basal joint longest and most robust, the second and third of nearly equal length, the third being, however, much slenderer than the second, the antennæ are terminated by a gradually formed slightly compressed club, the wings are almost diaphanous, surrounded by a distinct nerve, the cilia being very short, and the discoidal cell in all the wings being closed, the fore wings are somewhat triangular, but with the apex and posterior angle rounded off, the apical nervure, which is the third anterior branch of the postcostal, is forked more strongly than in the garden whites, the legs and ungues are formed as in the same insects The larva is elongated, slightly fusiform, hirsute, and the chrysalis is angulated, but not boat-shaped, with an obtuse beak in front of the tail, conical, attached in the same manner as all the other Pierides

The palpi have been relied upon as the chief character of the genus, but from the variations which are found even in the British species of garden whites, it appears to me an unsatisfactory one The antennæ and subdiaphanous wings are of more importance

I must refer to my observations under Pieris for the reasons which have induced me to reject the name of Pieris used for this insect by Stephens and others, and to employ one proposed long ago for it by Hubner It is on this account that I have also rejected Donzel's name, Leuconea†

* Probably derived from the Greek ἀπορία, destitutus sum from the nakedness of the wings

† M Donzel was induced to form this insect into a genus intermediate between Pieris and Parnassius, chiefly from the peculiarity which he observed, that the female carries the male when flying together coupled He further states that it differs from Pieris in having ten distinct terminal nerves, whilst there are only nine in Pieris he, however, overlooks the small apical branch which also exists in Pieris, although it is sometimes necessary to denude the wing of scales before it can be seen in that genus

SPECIES 1.—APORIA CRATÆGI, *Hubner*. THE BLACK-VEINED WHITE BUTTERFLY.

Plate vii. fig. 5—8.

SYNONYMES.—*Papilio (Dan. Cand.) Cratægi*, Linn. Syst. Nat. ii. 758. Lewin Papil. pl. 24. Donovan Brit. Ins. vol. xiii. pl. 454. Albin Brit. Ins. pl. 2. fig. 2 a—d. Wilkes M. & B., pl. 95. Harris Aurelian, pl. 9, fig. g—k.

Pieris Cratægi. Schrank, Latreille, Boisduval, Zetterstedt, Stephens, Curtis B. E. pl. 360. Duncan Brit. Butt. pl. 11, fig. 2.

Pontia Cratægi, Fabr., Ochs., Leach.

Leuconea Cratægi, Donzel, Ann. Soc. Ent. de France, 1837, p. 80.

Aporia Cratægi, Hubner, Verz. bek. Schmett.

This remarkable insect varies in the expanse of its wings from $2\frac{1}{3}$ to $2\frac{5}{6}$ inches. The wings are entirely of a white cream-colour, and are alike on both sides, the veins being black and more or less dilated, their extremities on the fore wings being accompanied by triangular dusky spots. In the female the veins of the fore wings are generally of a brownish hue.

The caterpillar is at first black, but is afterwards thickly clothed with whitish hairs, with the sides and belly of a leaden grey colour, marked with two longitudinal red or yellow stripes. The chrysalis is of a greenish white, with two lateral yellow lines and a great number of black dots.

This is a very destructive insect on the Continent, its larva feeding in society under a silken web not only on the white thorn (Cratægus oxyacantha), but also on the Prunus spinosa, the cherry-pear, and other fruit trees. M. Kollar has given a long and interesting account of its proceedings in his work on obnoxious insects, to which I must refer the reader, and of which a translation illustrated with woodcuts has recently been published by the publisher of this work. De Geer has also given an account of its transformations. It is fortunate, however, that this insect is of uncommon occurrence in this country, so that hitherto we have not experienced any of the injuries which it is capable of inflicting, and which led Linnæus to call it the pest of gardens. Pallas also relates in his travels that he saw this butterfly flying in such vast abundance in the environs of Winofka, that he at first took them for flakes of snow. It appears in this country somewhat periodically, being found plentifully in the New Forest in Hampshire, and at Combe wood in Surrey, although I have never seen it in the latter place during many years' collecting. It has also been taken at Chelsea, Muswell Hill, Herne Bay, Glanville's Wooton, Dorset, Enborne, Berkshire, and other parts of the south of England. In France there appear two broods, one in the spring, the other in autumn. Their periodical appearance may probably be owing to the failure in the preceding year of their natural enemies, but the cultivator ought to take advantage of their appearance in the winged state in order to prevent their increase, as the destruction of one female butterfly would prevent the deposition of a certain number of eggs, and the mischief attendant thereupon.

FAMILY II.

NYMPHALIDÆ, SWAINSON.

This, which is the third family of the butterflies, is in effect the second British family, there being no indigenous species of the second family Heliconidæ. The butterflies of which it is composed are for the most part very beautifully coloured, and of very robust structure, so that their flight is powerful and quick. They may be said to be of the middle size, few equalling the giant size of some of the Papilionidæ. Their chief characteristics, however, consist in the very short fore feet in both sexes which are quite unfitted for walking, and in the chrysalides being simply suspended by the tail. The antennæ have the extremities generally furnished with a more or less distinct club, which is never hooked; the hind legs have only a simple pair of spurs at the extremity of the tibiæ, and the posterior wings have a groove to receive the abdomen. There is scarcely any variation in the arrangement of the veins of the wings throughout the genera of this family, if we except the more or less complete closing of the discoidal cell of the hind wings. The caterpillars are variable in their structure, but in general they are clothed with numerous strong spines, others have the body smooth with the head or tail forked. The chrysalides are naked, and often armed with small conical protuberances. They are also often ornamented with golden or silvery spots. It will be observed that, by the arrangement of the butterflies here adopted, the Papilionidæ are far removed from the Lycænidæ, which agree together in the girthed condition of the chrysalis, and in the fore feet being fitted like the others for walking. M. Boisduval has endeavoured to obviate this objection by introducing the Lycænidæ between the Papilionidæ and Heliconidæ, whilst Dr. Horsfield has commenced the arrangement of the butterflies with the Lycænidæ followed by the Papilionidæ. As, however, I consider the Papilionidæ as the types of the Diurnal Lepidoptera, and consequently as most worthy to be placed at the head of the section, and as there certainly exists a natural transition from the Papilionidæ to the Heliconidæ and Nymphalidæ (see my Introduction to Mod. Classif. of Insects, v. 2, p. 342—353), I have adopted the arrangement of Stephens and other English authors. Some of the genera of this family recede from the others in having the club of the antennæ slender and very gradually formed, the larvæ smooth, with an anal fork and the pupa smooth. These have been separated by Dr. Horsfield as one of the five primary groups of butterflies, but the genera thus characterised, Apatura, Hipparchia, &c., possess so few characters in common, and are in other respects so closely allied to the typical Nymphalidæ, that it is not material, in a work of such confined limits as the present, to separate them therefrom.

GENUS IX.

MELITÆA*, Fabricius.

This genus is distinguished from the majority of the family by the very large and flattened club of the antennæ and the naked eyes. The palpi are long, ascending, and wider apart at the tip than at the base. The second joint is by far the largest, the third joint being small but variable in shape. The head is of moderate size; the fore wings rather long and triangular, but with the outer margin always rounded; the hind wings are rounded, and generally destitute of silvery markings. The fore legs are spurious in both sexes†; the four hind legs are terminated by tarsi, described as having double nails, or with simple claws furnished with an unguiform

* A fanciful name, probably derived from μελι honey, from the ground colour of the wings; or perhaps from Μελιταῖα the name of an ancient town in Thessaly.

† Mr. Curtis describes them as similar in the sexes, imperfect, hairy, and four or five jointed; his pl. 386, fig. 8 represents the fore leg as very hairy, and 8 b the tarsus as composed of three joints. The anterior tarsi, however, offer a most tangible character for the determination of the sexes. Their structure, now for the first time described, is as follows:—In the males they are not only very much more hairy than in the females, as pointed out by Zetterstedt, but entirely destitute of articulations, whilst in the females they are much less hairy, and distinctly composed of five joints, even without denuding them of scales, each of the joints having two short spines at the extremity on the inside. Mr. Curtis's description is, therefore, that of the female, and his fig. 8 that of the male fore leg; but his fig. 8 b must, I apprehend, be erroneous.

appendage, the ungues are, however, simple, acute, and strongly curved, each with an external pubescent, curved and bifid appendage on the outside, and there is a large fleshy pulvillus between the ungues. The characters laid down by Entomologists for the separation of this genus from the next, appear to me by no means satisfactory, indeed, the French and Germans seem to rely chiefly on the presence or absence of silvery spots on the under side of the wings, they accordingly unite Euphrosyne and Selene with Argynnis, of which, indeed, they are made the types both by Hubner and Ochsenheimer, which is thus made to contain very distinct groups. The fact, however, appears to me to be that the Fritillaries, as these spotted butterflies are called, instead of forming two genera, constitute a number of sub-genera of equal rank. With this view I propose the following arrangement, upon external characters alone, of the British Fritillaries.

1. Fore wings with the anterior margin straight or slightly concave, exterior margin rounded. Artemis.

2. Fore wings with the anterior and exterior margins rounded, not silvery beneath. Athalia, Cinxia.

3. Anterior and exterior margins of the fore wings rounded, hind wings silvered beneath. Euphrosyne, Selene.

4. Fore wings broad with simple veins, the fore margin rounded, and the outer margin concave. Lathonia.

5. Fore wings broad with dilated discoidal veins in the males, and with the outer margin generally concave. Paphia, &c.

This distribution appears to me to be confirmed by the structure of the palpi, thus, Artemis and Cinxia materially differ from each other in this respect although Mr Stephens places them in the same subsection. As, however, our English species constitute two, at first sight tolerably distinct, groups, founded on a general uniformity and smallness of size, and rounded outer margin of the fore wings in Melitæa, and a larger size, generally accompanied by a concave outer margin to the fore wings in Argynnis, I shall adopt the arrangement of our English authors.

DESCRIPTION OF PLATE VIII

INSECTS.—Fig 1 Melitæa Cinxia (the Glanville Fritillary Butterfly) 2 Showing the under side 3 The Caterpillar 4 The Chrysalis

" Fig 5 Melitæa Artemis (the greasy Fritillary B.) 6 Showing the under side 7 The Caterpillar 8 The Chrysalis

" Fig 9 Melitæa Athalia (the pearl bordered likeness B.) 10 Showing the under side 11 The Caterpillar 12 The Chrysalis

" Fig 13 Melitæa Pyronia (a variety of M Athalia) 14 Showing the under side

PLANTS.—Fig 15 Veronica Chamædrys (speedwell) 16 Scabiosa succisa (devil's-bit, scabious) 17 Lattuna vulgaris (common heath)

" Fig 18 Plantago lanceolata (ribwort)

The plan of this work is to give as far as possible an entire genus upon each plate, so as to present the points of difference of each species in juxta position, without having to refer backwards and forwards, during which operation the eye can scarcely carry minute differences with sufficient accuracy for the purposes of comparison. The advantage of this arrangement is particularly shown in the present plate, containing a large portion of the genus Melitæa. The three species, Cinxia, Athalia, and Artemis, which occurring in successive pages would appear to the inexperienced eye precisely similar are now, placed side by side, made clearly to display their differences of marking. Cinxia, with its range of black spots upon the border of the posterior wings, Artemis, with a somewhat similar range of spots, but varied by a central band of markings of a lighter colour upon both anterior and posterior wings, and Athalia deprived entirely of the range of spots, thus at once appear, even without reference to the under sides, so distinct as to enable the most unpractised eye to distinguish them at a glance. The differences of the caterpillars are still more slight, but it will be perceived that that of Cinxia has the head and legs brown, in that of Artemis the legs only are brown, and it has the additional distinction of a row of white markings along the side, while in that of Athalia, the legs and head are black and the spines only are brown. The caterpillars of Cinxia and Athalia are from Godart, that of Artemis from Hubner. H. N. H.

SPECIES 1.—MELITÆA ARTEMIS, THE GREASY FRITILLARY

Plate viii fig 5—8

SYNONYMS.—*Papilio Artemis*, Fabricius, Lewin Brit Papil pl 15, Harris Aurel pl 28 fig e—i
Melitæa Artemis, Fabricius, Ochsenheimer, Hubner, Stephens, Duncan Brit Butt pl 13, fig 2

Papilio Maturna, Esper
Papilio Iole, Borkhausen
Papilio Matutina, Thunberg
Papilio Lucina, Wilkes Eng Ins pl 111

This very distinct species varies in the expanse of its wings from 1½ to nearly 2 inches The wings on the upper side are of a darkish orange colour, varied especially from the base to beyond the middle of the wing with black and straw-coloured markings, which are succeeded on both wings by a broad orange bar, ornamented in the fore wings of fine specimens with a row of small straw-coloured dots, and on the hind wings with a similar row of black dots, this bar is followed by a row of black lunules, between which and the black margin of the wing is a row of orange or straw-coloured spots The fore wings beneath are of a paler and more obscure orange colour, with the black and straw-coloured marks almost obliterated, except at the tip, which is pale The hind wings beneath have three yellowish bands margined with thin black lines, the first near the base, irregular and oblique, the second broader and curved in the middle of the wing, and the third composed of marginal lunules, between which and the preceding bar is a row of black dots each surrounded with a pale circle The specimens vary considerably in the intensity and size of the markings One of these varieties is figured by Mr Duncan in his British Butterflies, plate 14, fig 2, but mistaken for Melitæa Cinxia Two others are figured by Mr Dale in Loudon's Mag of Nat Hist No 34 The fore wings have the anterior margin straight or rather slightly concave, the palpi are comparatively short, with the terminal joint nearly half as long as the preceding, and attenuated to the tip

The caterpillar is very spiny, black above, and yellowish beneath, with a row of small white dots down the back and sides, the legs are red-brown, the head and spines of the body black It feeds on the Devil's-bit Scabious, and both the species of Plantain When full grown it draws several blades of grass together and fastens them at the top with threads, suspending itself, according to Moses Harris, in the centre beneath The chrysalis is pale with dark spots The caterpillars are hatched in the autumn, the young brood passing the winter under a common web They are full fed in April The chrysalis state continues about a fortnight, and the butterfly is found in the months of May, June, and July, in swampy places, and is thence called the Marsh Fritillary by Bingley The following localities have been given for it Near Brighton, Enborne, Berks, Beachamwell Norfolk, Clapham Park, Bedfordshire, Glanville's Wootton and Dartmoor, Ambleside, Monk's Wood, Huntingdonshire, Eriswell, Mildenhall, and near Beccles, Suffolk, near Haverfordwest, near Durham, and near Belford, Northumberland Near "Coleshill, Woodstock, and Coventry" (Rev W T Bree)

PAPILIO MATURNA of Linnæus, a species closely allied to Artemis, but differing especially in the dark markings of the under side of the fore wings, and the want of the row of black dots near the margin of the hind ones, has been recorded by Stewart and others as a native species but on insufficient authority, having been mistaken for P Maturna, Fabricius, or M Athalia

SPECIES 2.—MELITÆA CINXIA. THE GLANVILLE FRITILLARY.

Plate viii. fig. 1—4.

SYNONYMES.—*Papilio Cinxia*, Linn. Lewin Pap. pl. 14. Donovan, pl. 242, fig. 1. Wilkes Brit. Ins. pl. 111. Harris Aurelian, pl. 16, fig. a—f.
Melitæa Cinxia, Ochsenheimer, Boisduval, Stephens, Curtis.
Schaenis Cinxia, Hubn. (Verz. bek. Schm.)
Papilio Delia, Hubner, Pupil.
Papilio Pilosellæ, Esper.
Papilio Trivia, Schrank.
Papilio Abbacus, Retzius.

This butterfly varies from $1\frac{1}{4}$ to nearly 2 inches in the expanse of the wings. The upper surface of the wings is uniformly fulvous, with numerous black markings; the base and costal edge of the fore wings is black, with fulvous dots; the discoidal cell has a broad central and apical black bar, with a fulvous middle; beyond this, there are four waved bands of black, the last being marginal; the same markings exist in the hind wings, besides which there is a row of round black dots on a fulvous ground near the margin of the wing. The fore wings, beneath, are fulvous, with the black spots nearly obliterated; the apex being very pale yellow with black dots; and the hind wings have three broad and very irregular pale yellow bars edged with thin black lines, and marked with black dots, between the basal and middle one of which is an irregular fulvous bar, and between the middle and apical ones is a row of fulvous eyes, with a black dot for a pupil. The fore wings have the costal margin evidently rounded; the palpi are long, with the last joint more than half the length of the second joint, and very acute at the tip. The caterpillar is intensely black, very slightly spotted with white; the head and prolegs fulvous. The chrysalis is brownish, with rows of fulvous tubercles on the back. The butterfly is rare. It appears in June and July (Harris says the middle of May). The caterpillars are hatched in the autumn, living in societies under a kind of tent formed by drawing together the tips of the leaves on which they feed, and covering them with a web. It is found in meadows near woods, especially in the South of England. Near Ryde and the Sandrock Hotel, Isle of Wight; Dover, Birchwood, Dartford, Kent. It has also occurred in Yorkshire and Lincolnshire. "Once taken by Mr. Walhouse near Leamington."—Rev. W. T. Bree.

SPECIES 3.—MELITÆA ATHALIA. THE PEARL-BORDERED LIKENESS FRITILLARY.

Plate viii. fig. 9—12.

SYNONYMES.—*Papilio Athalia*, Esper.
Melitæa Athalia, Ochsenheimer, Stephens, Duncan Brit. Butt. pl. 12, fig. 2.
Cinclidia Athalia, Hubner, (Verz. bek. Schmet.)
Papilio Dictynna, Lewin, pl. 11, fig. 5, 6. Haworth, Jermyn.
Papilio Maturna, Fabricius, Wilkes, pl. 112. (The Heath Fritillary.) Harris Aurel. pl. 28, fig. f, g.
The White May Fritillary, Petiver.

This species is nearly an inch and three quarters in expanse. The upper surface of the wings is black, the fore wings having two fulvous bars across, and one at the extremity of, the discoidal cell, and with a narrow bar behind the middle of the preceding; there are also three curved rows of fulvous spots between the middle and external margin of the wing, the third of which has the markings much smaller than the two preceding bars; these markings also run across the hind wings, the base of which is black, with a very few small spots. Beneath, the fore wings are fulvous, with slight black marks indicating the situation of the principal markings on the upper side; the spots along the outer margin and at the apex are straw-coloured. The hind wings beneath are fulvous at the base, which colour extends nearly to the middle of the wings; nearly at the base is an irregular bar of four straw coloured spots, succeeded by a single spot of the same colour, all having a slender edge of black; a broad curved bar of straw-colour runs nearly across the middle of the wing, margined with black lines, and having a slender black line running irregularly through it towards the base; this bar is succeeded by a row of fulvous lunules, the margin of the wing consisting also of straw-coloured lunules edged with black, and having a thin scolloped line of black running through it close to the margin.

Several varieties of this species are described by Mr Stephens, varying in the size of the fulvous markings, whereby they either become confluent from their larger size, or are almost obliterated by the black becoming more prominent.

The caterpillar is black and spiny, with two white dotted lines on each segment, and white tubercles on the sides. It feeds on the narrow and broad-leaved plantain, and according to Wilkes on the common heath. The butterfly appears from the beginning of May to July, frequenting heaths, marshes, &c. It is rare near London, but abundant in some parts of Devonshire, Dartmoor, and near Bedford, Coombe Wood, Hartley Wood, Essex, Apsley Wood, Bedfordshire, Caen Wood, Middlesex, and Faversham.

Melitæa Pyronia, Hubner, Stephens, or the Papilio Eos, Haworth, is considered by Ochsenheimer, Curtis, and Stephens, to be a variety of this species. That it is not a distinct species I infer from the irregularity and want of tessellation in the markings, the typical individuals of all the fritillaries being more or less distinctly tessellated. Our figures, plate 8, fig 13, 14, are copied from representations of this beautiful insect given by Mr Stephens (Illust Haust pl 4*, fig 1, 2). The specimen figured was taken by Mr Howard at Peckham in June 1803, and is rather more than an inch and a half in expanse, with the fore wings above deep fulvous, the veins, blotches in the middle, a waved streak and marginal band black, hind wings above black, with a waved bar of six fulvous spots beyond the middle, beneath, the fore wings are fulvous, but paler at the tips, with two black spots at the base, and a broad black bar in the middle divided by fulvous veins, and with a row of black lunules near the margin, hind wings fulvous at base, with about eight confluent black patches, the middle of the wing occupied by a broad whitish band, intersected by blackish veins followed by a row of fulvous lunules with black edgings, the outer margin straw-coloured, with a row of ochraceous lunules in the middle.

Papilio Tessellata, *scrotina, subtus stramínea*, or the Straw May Fritillary of Petiver (Papil pl 3, fig 11, 12, and our plate 9, fig 13, 14), is also now considered by Stephens and Curtis as a variety of P Athalia. In Petiver's time it was "pretty common in Caen Wood," where Athalia also occurred. It is paler above than that species and the fore wings are more fulvous beneath, the hind wings beneath are entirely straw-coloured, with black veins, at the base are three large yellow spots edged with black, a broad curved fascia of straw-yellow runs across the middle of the wings, edged with black, and with an irregular black line running through the middle of it, this is succeeded by a row of black lunules, and the margin is straw-yellow with a black vandyked line running along it.

DESCRIPTION OF PLATE IX

Insects.—Fig 1 Melitæa Selene (small pearl-bordered Fritillary Butterfly) 2 Showing the under side 3 The Caterpillar 4 The Chrysalis

" Fig 5 Melitæa Dia (the small purple Fritillary B) 6 Showing the under side 7 The Caterpillar

" Fig 8 Melitæa Euphrosyne (the pearl bordered Fritillary B) 9 Showing the under side 10 The Caterpillar

" Fig 11 A dark variety of M Euphrosyne

" Fig 12 A variety of M Euphrosyne, with pale under side (The Thalia of Haworth)

" Fig 13 Melitæa Tessellata 14 Showing the under side

Plants.—Fig 15 Viola Tricolor (Heartsease) 16 Viola canina (Dog's violet)

Two of the insects figured in this plate, M Selene and M Euphrosyne, are so much alike on the upper side, that it is difficult in a drawing to render them very distinct. The two individuals from which the drawings were made were indeed somewhat different in colour, but that is no[illegible]

constant difference, and can therefore be no guide The markings, however on the posterior wings are somewhat different, those of Selene being of a rather more regular character, and the dot near the base round and distinct, whilst in Euphrosyne it is almost always shaded off indistinctly into the dark colour at the base of the wing I have noticed also, and endeavoured to have it expressed in the plate, that the rings of white upon the antennæ are broader in Selene, giving them a lighter and more checkered appearance But though there is some difficulty in distinguishing the insect on the upper side, it will be seen that the under sides are very dissimilar Both specimens are from the collection of Mr Westwood The Caterpillars are from Hubner The dark variety of Euphrosyne, which on the under side is the same as in ordinary specimens, is from the collection of Mr Westwood, and the pale variety is the one figured by Mr Stephens Melitæa Dia, the petite violette of the French, so called from the beautiful purple blush that suffuses the posterior wings on the under side, is from the specimen in the collection of Mr Stephens, and the Caterpillar likewise from Hubner

Melitæa Tessellata is from the old and somewhat coarse etching of Petiver, and perhaps does not convey a very accurate idea of the insect, but I have been induced to give it, in order that this and the preceding plate might contain all the small Fritillaries reputed British H N H

SPECIES 4 —MELITÆA SELENE THE APRIL OR SMALL PEARL-BORDERED FRITILLARY

Plate ix fig 1—4

SYNONYMES —*Papilio Selene* Fabricius *P Selene* Haworth *Melitæa Selene*, Stephens Curtis Brit Ent pl 386 Duncan Brit Butt pl 10, fig 3 (*M Selene*) *Argynnis Selene*, Ochsenheimer, Boisduval, Hubner (Verz) *Papilio Euphrasia*, Lewin Pap pl 13 *Papilio Euphrosyne*, var Esper Harris Aurelian pl 31, fig i-k

This species varies in its expanse from 1- to nearly 2 inches It is fulvous on the upper side, spotted with black, four irregular bars run across the discoidal cell, succeeded first by a single spot extending from the costa, and then an interrupted row of four spots, then a row of seven small round black dots and a row of black lunules reaching to the margin, which is edged with black, the hind wings are similarly marked beyond the middle, but the base is black, with numerous fulvous angular shaped marks, amongst which a spot on the centre of the discoidal cell, with a round black dot in the middle, is most conspicuous The fore wings on the under side are nearly marked as above, except that the spots are smaller, and the apex is varied with ferruginous and straw-coloured, the hind wings beneath are most beautifully tesselated with white, straw-colour, buff, dark ferruginous and silver, the several markings being edged with black lines and the veins being black The red mark in the middle of the discoidal cell with a black dot in the centre, is here conspicuous, succeeded by an oblong silvery patch, between which and the outer angle are three smaller silvery patches, there being also a marginal row of six wedge-shaped silvery marks on the outer margin The hind wings, as shown in our figures 2 and 9, are much narrower than in M Euphrosyne, so that the space between the longitudinal veins is narrower, and the markings consequently not so broad

The caterpillar is black, with a pale lateral stripe, the spines are half yellow, and two on the neck are larger than the rest, and project forwards The chrysalis is dirty greyish-coloured

This species is common in various places in the South of England, Dartmoor, Lyndhurst, Newcastle, and Durham, have also been mentioned as its localities It frequents heaths and waste grounds Although occasionally captured in May and July, the beginning of June appears to be the period for the exclusion of the first brood, the second being produced in August and September *.

It is liable to vary considerably, Mr Stephens describes a specimen with the upper surface of the wings whitish

---

* In London's Mag Nat Hist , No 21, are some observations by the Rev T W Bree relative to the double broodedness of this species in reply to which, Mr Newman stated, in No 22, that this species appears in the summer fifteen days later than M Euphrosyne, and lasts till the end of July, after which it never reappears Mr Dale, however (Ent Mag 1 357), speaks of it as double brooded, and that the two broods vary in the same manner as those of M Euphrosyne

PAPILIO THALIA of Hubner and of Haworth, is also considered by Mr Stephens as an accidental variety of this species, although it has been questioned whether the Thalia of Haworth may not be a variety of Euphrosyne The latter variety is copied, in our plate 9, fig 12, from Mr Stephens's figure drawn from Mr Haworth's specimen, and is thus described by Mr Stephens —"Wings above, pale fulvous, irregularly spotted with black, anterior, beneath, pale, varied with yellowish and ferruginous towards the tips, with some obsolete black and dusky spots on the disc, posterior wings variegated with ferruginous yellowish and greenish, with the pupil of the ocellus very large, the discoidal silvery spot produced to the hinder margin, and the usual marginal spots lengthened inwardly, the usual fasciæ are obliterated, but the silvery spot at the base is somewhat apparent"

SPECIES 5—MELITÆA EUPHROSYNE THE PEARL-BORDERED FRITILLARY

Plate ix fig 8—10

SYNONYMES—*Papilio Euphrosyne*, Linnæus Lewin Pap pl 13 Donovan pl 312
Melitæa Euphrosyne Leach, Stephens Curtis, Duncan Brit Butt pl 15, fig 2
Argynnis Euphrosyne, Ochsenheimer, Boisduval, Hubner (Verz) Harris Aurelian, pl 40, fig e, f

This species is closely allied to the last but is rather larger, varying from 1⅔ to nearly two inches, and having the hind wings far less strikingly variegated on the under side The upper side of all the wings so closely resembles those of M Selene, that no further description is required of them, the under side of the fore wings is also similar to that of the same species, but the black markings are not so distinct, and the apex of the wing has the buff much deeper and the ferruginous marks much paler The hind wings, beneath, have all the markings much less distinct than in Selene, there being moreover only one small patch towards the base, a large spot at the apex of the discoidal cell, and seven marginal wedge-shaped marks of silver The centre of the discoidal cell is rusty red, with a yellowish spot in the centre, having a black dot in the middle, between the central and marginal silvery spots is a row of round rusty-red dots

Mr Stephens mentions several varieties of this species, in one of which the silvery marginal spots are wanting, another, with "the basal half of all the wings, above, black, spotted with fulvous with large black spots on the anterior wings beneath," seems in some degree to resemble the specimen figured in our plate 9, fig 11, from my collection in which all the black markings on the upper side of the fore wings are suffused, except the row of submarginal round spots, the markings on the hind wings are somewhat more distinct The underside scarcely differs from typical individuals The Rev Mr Bird possesses another variety nearly white

This species is the most abundant of all the Fritillaries, especially in woods in the southern parts of the kingdom, it is also found plentifully in various parts of Scotland The larva is black and spiny, with two rows of orange dots on the back

It feeds on various kinds of violet, and there are two broods in the year, the butterfly first appearing in May and again at the beginning of autumn*

* The Rev W T Bree published some observations relative to the double-broodedness of this species in Loudon's Mag of Nat Hist No 21, in reply to which, Mr Newman, in the next number of the same work, stated that at Birch Wood, Kent, this butterfly appears at the end of May by thousands, and lasts till the end of June, but that it never reappears afterwards Mr Dale, however (Ent Mag 1 357), speaks of it as double-brooded, and states that the spring brood varies very much in its markings, and that the September brood varies in colour, being much yellower

SPECIES 6.—MELITÆA DIA.

Plate ix. fig. 5—7.

SYNONYMES.—*Papilio Dia*, Linnæus, Stewart, Turton.
Melitæa Dia, Stephens, Jermyn. Loudon's Mag. Nat. Hist. vol. v. 751, fig. 121.
Argynnis Dia, Ochsenheimer, Hubner. (Verz.)

This species which is also closely allied to M. Selene, is about an inch and three quarters in expanse. The upper surface of both pair of wings has the black spots and markings larger and stronger, the base of the posterior being nearly black, so that the whole assumes a darker appearance than in M. Selene. But the principal difference consists in the under side of the posterior wings, which are of a brownish purple interspersed with darker markings of the same colour, and numerous irregular semi-metallic spots, of which there are six or seven of small size at the base, intermixed with minute yellowish dots, a band composed of silvery and yellowish spots, then a purplish white streak, in which is a series of circular spots slightly pupillated, and finally, in the margin is a series of silvery lunules.

The caterpillar is black, with the spines white and reddish, the back greyish, with a longitudinal line. It feeds on the Viola odorata, and there are two broods in the year.

Found by Mr. Weaver several times in Sutton Park, near Birmingham, and also near Alderley, in Cheshire, by Mr. Stanley.

GENUS X.

ARGYNNIS*, FABRICIUS.

Referring to the observations under the genus Melitæa relative to the characters of that and the present genus, we may define this to be distinguished chiefly by the larger size of the insects, the silvery spots which ornament the under side of the wings, which are broad and of ample size, the ordinarily concave posterior margin of the fore wings, the tessellated appearance of their upper surface, and the dilatation of the branches of the median and the anal vein of the fore wings in the males of most of the species. The antennæ are terminated by a suddenly-formed broad compressed, or rather spoon-shaped, club, the head is broad, the eyes are large and naked, the fore legs rudimental.† The ungues of the four posterior tarsi are formed as in Melitæa, and their structure has been carefully illustrated in the Crochard Edition of the Regne Animal. Insects, plate 135. I have purposely omitted all mention of the form of the palpi in the above characters, as this character does not appear to me of any value in separating the Fritillary butterflies into two genera, the true types of Melitæa, or those without silvery spots, having the terminal joint as large and acute as it is in the typical Argynnes, whilst Mr. Stephens has observed that Lathonia and Euphrosyne agree together in their palpi. Lathonia moreover differs from the other Argynnes in several other important respects, so that it must evidently be regarded as an intermediate form. I therefore place it at the head of the genus in order that it may be brought into connexion with the silvery spotted Melitææ.

* A fanciful name being one of the denominations of Venus (Vollm. *Vollst. Worterb. der Mythol.* Stuttg. 1836).

† The fore legs are described by Curtis as "alike in both sexes." They differ, however, in the sexes, in the same manner as the fore legs of Melitæa described in a preceding page. I have represented their structure in P. Paphia, in my Introd. to Mod. Classific. vol. 2. p. 353, fig. 95, 4, 5, 6, 7.

DESCRIPTION OF PLATE XI

INSECTS.—Fig. 1. Argynnis Adippe, female (the high brown Fritillary Butterfly). 2. Showing the under side. 3. The Caterpillar.

" Fig. 4. Argynnis Lathonia (the Queen of Spain Fritillary, B.). 5. Showing the underside. 6. The Caterpillar. 7. The Chrysalis.

PLANTS.—Fig. 9. Anchusa officinalis (the Alkanet). 10. Onobrychis sativus (the Common Sainfoin).

I have not thought it worth while to give both male and female of A. Adippe, as they are very similar, the male being merely a little smaller, of a somewhat richer colour, and having two of the nervures of the anterior wings slightly thickened. At first sight, the insect appears very like A. Aglaia, but, upon examination, it will be found that the markings are much more open; and on the under side the difference is rendered more distinct by the brown ocellated spots of the posterior wings, in which it more nearly resembles A. Lathonia. A. Lathonia is, however, very distinct from every other species of this genus, both by the fine regular spotting of the wings on the upper side, and also by the profusion of metallic marks on the under side of the posterior wings, which in some brilliant varieties taken on the Continent, form one united plate of silver. A. Lathonia is from a specimen in my own possession, A. Adippe from the collection of Mr. Westwood, and the Caterpillars are from Godart. H. N. H.

SPECIES 1.—ARGYNNIS LATHONIA. THE QUEEN OF SPAIN FRITILLARY.

Plate xi. fig. 4—7.

SYNONYMES.—*Papilio Lathonia*, Linnæus, Lewin pl. 12. Donovan Brit. Ins. vol. iii., pl. 73.
Argynnis Lathonia, Fabricius, Ochsenheimer, Leach, Stephens, Curtis, Duncan Brit. Butt. pl. 16, fig. 2.
Issoria Lathonia, Hubner (*Verz.*)
Papilio Primipissa, Linnæus, olim.
Papilio Latonia, Denn. and Schiff. *Argynnis Latonia*, Zetterstedt.
Papilio Lathona, Hubner (*Pap.*)

This exquisite insect is generally about two inches in expanse. The upper surface of the wings is fulvous-orange, with numerous very distinct and mostly rounded black spots, those of the apex of the fore wings uniting with the dark margin and inclosing several small paler buff patches. The anal and median veins are not dilated in the males. Beneath, the fore wings are marked nearly as above, except that the apex of the wing has a broad ferruginous patch, at the base of which is a silvery spot, succeeded by two small eyes, between which and the margin are several oval silver patches; the hind wings on this side are pale buff varied with reddish brown, ornamented with numerous silvery patches, varying greatly in size and form, of which there are about fourteen between the base of the wings, and a row of seven dark brown ocelli with silvery pupils, between each of which and the margin of the wing is a large silvery patch.

The caterpillar, according to Godart, is greyish-brown, with a white dorsal line spotted with black, and with two brownish-yellow lines on the sides; the spines and legs pale yellow; it feeds on heartsease, sainfoin, and borage. The pupa is varied with brownish and greenish, and ornamented with metallic spots. The perfect insect appears in August and September; but, according to Godart, the later specimens survive the winter, and again appear in Spring. Mr. Dale says there are two broods (Ent. Mag. i. 356). By Petiver it is recorded as occurring again in May; but Mr. Stephens' specimens captured in the middle of August were much faded, so that he is led to believe that the species is double-brooded. This butterfly, although still accounted a great rarity, occurs in numerous situations wide apart. The following are some of its localities:—Gamlingay, Cambridgeshire; Stoke-by-Nayland; near Wisbeach; Halvergate, Norfolk; Battersea Fields; Dover; Colchester; Birch Wood, Kent; Hertford.

SPECIES 2.—ARGYNNIS ADIPPE. THE HIGH-BROWN FRITILLARY.

Plate xi. 1—3.

SYNONYMES.—*Papilio Adippe*, Linnæus, Esper, Lewin's P. pl. 30. Donovan Brit. Ins. pl. 418. Harris's Aurelian pl. 28, fig. a—d.
Argynnis Adippe, Fabricius, Ochsenheimer, Stephens, Duncan Brit. Butt. pl. 16, fig. 1.
Acidalia Adippe, Hubner (Verz.)

This species varies in the expanse of the wings from 2½ to 2¾ inches. The upper surface is uniformly of a rich fulvous-orange, except at the base, which is greenish, with numerous black markings, many of which assume a crescent shape, especially the row running near the outer margin of the wing, and which are united to two slender marginal black lines The fore wings in the males have the two inner branches of the median veins strongly dilated in the middle, the anal vein being scarcely dilated The under side of the fore wings almost resembles the upper, except that the spots towards the apex of the wing assume a rich brown colour, some being marked with silver spots The hind wings are on this side varied with buff, ferruginous, and brown, at the base are about seven silvery patches, disposed somewhat in a circle, beyond which is an irregularly curved row of about nine or ten silvery patches varying in size, succeeded by a row of rusty red spots, some of which have the centre silvery, there is also a row of seven submarginal silvery wedge-shaped spots In both sexes the outer edge of the fore wings is slightly concave

Several varieties of this species have been observed, in which the spottings of the wings become more or less confluent

The caterpillar is at first red, but subsequently olive-green, with a white line down the back, and white spots on the sides It feeds on the heartsease and sweet violet The chrysalis is reddish, with silvery spots This state lasts about a fortnight the butterfly appears at the end of June or beginning of July The butterfly frequents heaths and the borders of woods, and is far from uncommon in most of the southern counties of England

Godart, Ochsenheimer, and Curtis, consider it doubtful whether this insect be the P Adippe of Linnæus considering that the A Niobe of Hubner, &c, is the true Adippe Professor Zetterstedt has, however, shown the correctness of the ordinary opinion respecting the names of this species The true Niobe, which Stewart gives as British, and of which Mr Dale possesses a specimen, which he obtained from the professedly indigenous collection of Dr Abbot, is indeed very similar to our common Adippe, but it is rather smaller, with the base of the wings above more dusky, and the posterior beneath much more strongly variegated with yellow (or rarely silver) spots but a more important character is the very slight incrassation of the veins of the fore wings of the males

DESCRIPTION OF PLATE X

INSECTS —Fig 1 Argynnis Paphia, male (the silver washed Fritillary Butterfly) 2 The female 3 Showing the under side
4 The Caterpillar 5 The Chrysalis
" Fig 6 Argynnis Aglaia male (the dark green Fritillary B) 7 The female 8 Showing the under side
9 The Caterpillar 10 The Chrysalis

PLANTS —Fig 11 11 Rubus Idæus (the Raspberry) 12 12 Viola odorata (the Sweet Violet)

This plate contains only two species, but as they are two of the most remarkable and distinct of the larger Fritillaries, I have thought it advantageous to give three figures of each, particularly as the males and females present such differences of marking and colour as might lead the young collector to consider them as different species The male of A Paphia, it will be seen, is of a much brighter colour than the female, and the nervures of the interior wings are thickened into broad black stripes not found in the female

The male of A Aglaia, on the contrary, is not so fine an insect as the female, which is not only larger but more strongly marked, and the rounded spots enclosed by the black crescents of the border of the posterior wings are of a lighter and brighter tone than the ground colour In the male these spots are the same as the ground-colour, which is generally somewhat richer than the female The male and female of both species are from specimens in the collection of Mr Westwood, and the Caterpillars are from Hubner H N H

(*To be completed in about Sixteen Numbers.*)

No. V.] [PRICE 2s. 6d.

BRITISH INSECTS AND THEIR TRANSFORMATIONS.

BRITISH BUTTERFLIES

AND

Their Transformations.

ARRANGED AND ILLUSTRATED IN A SERIES OF PLATES

BY H. N. HUMPHREYS, ESQ.

WITH CHARACTERS AND DESCRIPTIONS

BY J. O. WESTWOOD, ESQ., F.L.S.,

SEC. OF THE ENTOMOLOGICAL SOCIETY, ETC. ETC.

LONDON:

WILLIAM SMITH, 113, FLEET STREET.

MDCCCXL.

WORKS PUBLISHED BY WILLIAM SMITH, 113, FLEET STREET.

SMITH'S STANDARD LIBRARY.

In medium 8vo, uniform with Byron's Works, &c

Works already Published.

THE LADY OF THE LAKE 1s
THE LAY OF THE LAST MINSTREL 1s
MARMION 1s 2d
THE VISION OF DON RODERICK BALLADS AND LYRICAL PIECES By SIR WALTER SCOTT 1s
THE BOROUGH By the REV G CRABBE 1s 4d
THE MUTINY OF THE BOUNTY 1s 4d
THE POETICAL WORKS OF H KIRKE WHITE 1s
THE POETICAL WORKS OF ROBERT BURNS 1s 6d
PAUL AND VIRGINIA, THE INDIAN COTTAGE, and ELIZABETH 1s 6d
MEMOIRS OF THE LIFE OF COLONEL HUTCHINSON, GOVERNOR OF NOTTINGHAM CASTLE during the Civil War By his Widow, Mrs LUCY HUTCHINSON 2s 6d
THOMSON'S SEASONS, AND CASTLE OF INDOLENCE 1s
LOCKE ON THE REASONABLENESS OF CHRISTIANITY 1s
GOLDSMITH'S POEMS AND PLAYS 1s 6d
———— VICAR OF WAKEFIELD 1s
———— CITIZEN OF THE WORLD 2s 6d
———— ESSAYS, &c 2s

⁂ These four Numbers form the Miscellaneous Works of Oliver Goldsmith, and may be had bound together in One Volume, cloth-lettered, price 8s

KNICKERBOCKER'S HISTORY OF NEW YORK 2s 3d
NATURE AND ART By MRS INCHBALD 10d
A SIMPLE STORY By MRS INCHBALD 2s
ANSON'S VOYAGE ROUND THE WORLD 2s 6d
THE LIFE OF BENVENUTO CELLINI 3s
SCHILLER'S TRAGEDIES THE PICCOLOMINI, and THE DEATH OF WALLENSTEIN 1s 8d
THE POETICAL WORKS OF GRAY & COLLINS 10d
HOME By MISS SEDGWICK 9d
THE LINWOODS By MISS SEDGWICK 2s 8d
THE LIFE AND OPINIONS OF TRISTRAM SHANDY By LAURENCE STERNE 3s
INCIDENTS OF TRAVEL IN EGYPT, ARABIA PETRÆA AND THE HOLY LAND By J L STEPHENS, Esq 2s 6d
INCIDENTS OF TRAVEL IN THE RUSSIAN AND TURKISH EMPIRES By J L STEPHENS Esq 2s 6d
BEATTIE'S POETICAL WORKS and BLAIR'S GRAVE 1s
THE LIFE AND ADVENTURES OF PETER WILKINS, A CORNISH MAN 2s 4d
UNDINE A MINIATURE ROMANCE Translated from the German, by the REV THOMAS TRACY 9d
THE ILIAD OF HOMER Translated by ALEX POPE 3s
ROBIN HOOD, a Collection of all the ancient Poems, Songs, and Ballads now extant, relative to that celebrated English Outlaw, to which are prefixed, Historical Anecdotes of his Life Carefully revised, from RITSON 2s 6d
THE LIVES OF DONNE, WOTTON, HOOKER, HERBERT, AND SANDERSON Written by IZAAK WALTON 2s 6d
THE LIFE OF PETRARCH By MRS DOBSON 3s
MILTON'S PARADISE LOST 1s 10d
———— PARADISE REGAINED, and MISCELLANEOUS POEMS 2s
RASSELAS By DR JOHNSON 9d
ESSAYS ON TASTE By the Rev ARCHIBALD ALISON, LLB 2s 6d
THE POETICAL WORKS OF JOHN KEATS 2s
THOMSON'S POEMS AND PLAYS, COMPLETE 5s

FOUR VOLUMES ARE NOW ARRANGED ON THE FOLLOWING PLAN —

ONE VOLUME OF "POETRY,"

CONTAINING

SCOTT'S LAY OF THE LAST MINSTREL
SCOTT'S LADY OF THE LAKE
SCOTT'S MARMION
CRABBE'S BOROUGH
THOMSON'S POETICAL WORKS
KIRKE WHITE'S POETICAL WORKS
BURNS'S POETICAL WORKS

A SECOND VOLUME OF "POETRY,"

CONTAINING

MILTON'S POETICAL WORKS
BEATTIE'S POETICAL WORKS
BLAIR'S POETICAL WORKS
GRAY'S POETICAL WORKS
COLLINS'S POETICAL WORKS
KEATS'S POETICAL WORKS
GOLDSMITH'S POETICAL WORKS

ONE VOLUME OF "FICTION,"

CONTAINING

NATURE AND ART By MRS INCHBALD
HOME By MISS SEDGWICK
KNICKERBOCKER'S HISTORY OF NEW YORK By WASHINGTON IRVING
PAUL AND VIRGINIA By ST PIERRE
THE INDIAN COTTAGE By ST PIERRE
ELIZABETH By MADAME COTTIN
THE VICAR OF WAKEFIELD By OLIVER GOLDSMITH
TRISTRAM SHANDY By LAURENCE STERNE

ONE VOLUME OF "VOYAGES AND TRAVELS,"

CONTAINING

ANSON'S VOYAGE ROUND THE WORLD
BLIGH'S MUTINY OF THE BOUNTY
STEPHENS'S TRAVELS IN EGYPT ARABIA PETRÆA, AND THE HOLY LAND
STEPHENS'S TRAVELS IN GREECE TURKEY, RUSSIA, AND POLAND

These Volumes may be had separately, very neatly bound in cloth, with the contents lettered on the back, price 10s 6d each

Those parties who have taken in the different works as they were published and who wish to bind them according to the above arrangement, may be supplied with the title-pages gratis, through their bookseller, and with the cloth covers at 1s each

SPECIES 3.—ARGYNNIS AGLAIA. THE DARK GREEN FRITILLARY.

Plate x fig 6—10

SYNONYMES.—*Papilio Aglaia*, Linnæus. Lewin Pap. pl. 11. Donovan pl. 302. Wilkes pl. 115. Harris Aurelian pl. 26, fig. o—p.

Argynnis Aglaia, Ochsenheimer, Stephens, Jermyn, Duncan Bri Butt. pl. 15, fig. 1.

Acidalia Aglaia, Hubner. (*Verz.*)

This species is closely allied to the preceding, which it very much resembles, especially on the upper side of the males, differing however in several characters which do not appear to have been previously attended to. The two inner branches of the median vein are much more slightly dilated in the males, the anal vein being on the contrary more strongly dilated; the outer margin of the fore wings in the males is almost straight, or scarcely perceptibly concave, whilst that of the females is distinctly rounded, and the hind wings are destitute beneath of the rich-coloured row of eyes between the two outer rows of silvery spots; it is, occasionally, also rather wider in expanse of the wings than A. Adippe. The general colour is also paler, with the marginal band darker-coloured. The females are much paler than the males, with the submarginal row of spots above still paler. Beneath, the hind wings are varied with green and yellow, with about seven silvery spots at the base, an irregular row of seven silver spots beyond the middle of the wing, and a row of seven submarginal transverse spots of silver, bordered above with greenish crescents.

PAPILIO CHARLOTTA of Haworth (Lep. Brit. p. 32, Sowerby Brit. Misc. pl. 11. Bree in London's Mag. of Nat. Hist. vol. v. p. 750, Arg. Carolotta, Jermyn), represented in our pl. 12, fig. 1, 2, is regarded by Stephens and Curtis as a variety of this species, differing from it in having two of the costal spots on both sides of the fore wings united, and only nineteen instead of twenty-one silvery spots on the under side of the hind wings, several of the ordinary spots at the base being confluent.

Several specimens of another still more striking variety (at first given by Stephens as a variety of Adippe) have also been captured, in which the upper surface of the fore wings is almost entirely of a dark brownish black, except a bright linear fulvous mark, and beyond it a much smaller mark of the same colour, with a row of faint tawny spots running parallel with the hinder margin. The hinder wings have the markings considerably more distinct. Beneath, the ground colour of the fore wings is dark ferruginous, and that of the hind wings pea-green, with twenty-one silvery spots. This variety has been figured by Curtis (Brit. Ent. pl. 290), and by the Rev. W. T. Bree (London's Mag. Nat. Hist. vol. v. p. 749), and has been taken near Ipswich and Birmingham. Mr. Curtis mentions a variety intermediate between this and the preceding; and in the Magazine of Natural History (No. 26), a pale buff-coloured variety is mentioned with the spots and markings very faint.

The caterpillar is blackish, with a whitish line down the back and another at the side, above which is a row of eight small red spots. It feeds on the dog's-violet. The perfect insect appears in July and August. It is a common species, and is found throughout the whole kingdom, frequenting heaths, meadows, woods, and downs.

DESCRIPTION OF PLATE XII.

INSECTS.—Fig. 1. Argynnis Charlotta (a variety of A. Aglaia). 2. Showing the under side.

,, Fig. 3. Argynnis Paphia (a beautiful female variety).

,, Fig. 4. Argynnis Aphrodite. 5. Showing the under side.

PLANTS.—Fig. 6. Viola lutea (the wild yellow heart's-ease).

As the Fritillaries are remarkable for accidental variations of colour and marking, I have devoted this plate to two of the most striking varieties known in England. A. Charlotta, which some consider a distinct species, appears something between A. Aglaia and A. Adippe. It however most strongly resembles Aglaia, particularly on the under side It differs from both in having one of the numeral figures on the costa of the anterior wings (which in the former species are said to resemble 1536) obliterated by its confluence with the next, and on the under side, the six basal silver spots of the posterior wings of Aglaia are in A Charlotta united into three larger ones These characteristic differences are constant in different individuals, which seems to entitle A Charlotta to rank at all events as a sub species Specimens taken recently by Mr Weaver at Sutton Park, and near York, agree perfectly in these characters The variety of A Paphia is from the specimen in the British Museum taken by Mr Dale forty years ago, which is I believe unique

A Aphrodite having been captured in England I have thought it well to give a figure of it also in this place, though I do not consider it a British insect Its having been captured at Upton Wood is no proof of its indigenous character a splendid Brazilian insect was recently caught in the conservatories of Messrs Loddiges of Hackney, imported no doubt in its preparatory state in a parcel of plants, and I know a fine collection of exotic beetles all caught alive in the London Docks H N H

SPECIES 4 —ARGYNNIS APHRODITE

Plate xii fig 1—5

SYNONYMES —*Papilio (Nymph) Aphrodite*, Fabr Ent Syst 3, p 144
Argynnis Aphrodite, Bree in Mag Nat Hist 2nd Ser vol iv p 131 Suppl Illust pl 10

An account of the capture of a specimen of this North American insect having recently appeared, we have added a figure of it, without, however, wishing thereby that it should be inferred that the species in question is a native of our island The specimen described by Mr Bree is represented in the plate above referred to as being nearly 3¼ inches in expanse It is described by him as being of a rich fulvous colour, chequered and spotted with black on the upper surface The black spots and markings on the second pair of wings are neither so large nor so strongly developed as in the corresponding wings of A Paphia, Aglaia, and Adippe, to which latter species it more nearly approaches on the under surface, having the second pair of wings adorned with numerous silver spots on a buff-coloured ground, which is dark towards the base of the wings and becomes lighter towards the outer extremities, with a marginal row of semicircular silver spots From Mr Bree's account, there appears to be no reason to doubt that the insect in question was taken by James Walhouse, Esq of Leamington, in Upton Wood, a few miles from that town, in the summer of 1833, and was presented to Mr Bree's son, in whose possession it now is, by Moreton J Walhouse, Esq , the brother of the captor Mr Bree considers that this specimen could neither have been imported from America nor have flown across the Atlantic It appears to me, however, not improbable that it might have been imported either in the egg or chrysalis state from America attached to American plants In a subsequent page of the same volume, Mr Bladon mentions his having seen a Fritillary near Pontypool, which he conjectures must have been this species

There are several very closely allied American species, including the present—if, indeed, they are not merely varieties of each other I mention this because the specimens of Aphrodite in the British Museum are larger than the figure above referred to

---

SPECIES 5 —ARGYNNIS PAPHIA THE SILVER-WASHED FRITILLARY

Plate x fig 1—5

SYNONYMES —*Papilio Paphia*, Linnæus Lewin Pap 9 Donovan 7, pl 217 Wilkes, pl 110 Harris Aurelian, pl 24, fig k—n *Argynnis Paphia*, Fabricius (type species) Ochsen , Stephens, Curtis Duncan Brit Butt pl 14, fig 1 *Argyronome Paphia*, Hubner (Verz d bek Schmett)

This is the largest of our strictly British Fritillaries, varying from 2¾ to 3 inches in the expanse of the wings, which are of a fulvous colour in the male on the upper side, but paler and tinged with greenish in the female, with numerous black spots and bars, there being three distinct rows of spots along the outer margin, the most external of which are diamond-shaped, besides which, the males have the anal vein and the three branches of the median vein strongly dilated and black in the middle. The under side of the fore wings is paler, with black marks, but those adjoining the outer margin are almost obliterated, and replaced near the tip with greenish scales. The hind wings are greenish, with two short silvery bars near the base, a narrower one running obliquely across the middle of the wing, and another marginal one, between the two last is a row of green circles, and another of green lunules forming the inner margin of the marginal band.

A fine variety of this species, which has been regarded as a distinct species and figured by Ernst, is in the British Museum, from which our plate 12, fig 3, was taken. It was captured many years ago by Mr Dale, and is a female, and has the upper surface of the wings very dark, with some whitish spots at the tips of the fore wings. Similar individuals have not unfrequently been met with on the Continent, where they are known under the ordinary name, "*le Valaisien*." Their specific identity with A. Paphia has been demonstrated in a remarkable manner—Hubner having figured (pl 190, fig 935, 936) a specimen apparently female, the right wings of which are coloured as in the variety and the left as in the type of the species. A still more remarkable specimen has been figured by M Wesmael in the 4th volume of the Bulletin of the Academy of Brussels, in which the right wings were those of the male type, except that the marginal row of spots were as large as in the female, the left fore wing exhibited a complete melange of the male and female, as well as of the variety and typical individuals, the ground colour being fulvous as in the male, but the markings, especially at the tip, dark as in the female, with the white spots of the variety, the upper side of the hind wings entirely coloured as in the variety.

Another gynandromorphous individual is mentioned by Ochsenheimer, the right wings of which are those of the male, and the left those of the female. In Loudon's Magazine of Natural History, the capture of an English specimen is noticed, according with Ochsenheimer's description.

The caterpillar is light brown, with a row of yellow spots on the back. The spines are long, the two next the head being longer than the rest. It feeds on the dog's-violet, raspberry, and nettle. The chrysalis is grey, with the tubercles gilt.

This is an abundant species especially in the south of England, occurring also in Scotland. It flies in July.

GENUS XI
VANESSA* Fabricius

This genus may be considered as comprising the most beautiful and highly ornamented of our British butterflies, distinguished generically from the preceding Nymphalidæ by having the eyes pubescent and the wings angulated, by which latter character, as well as by the more sudden formation of the club, they are separated from the terminal genera of the family. The head is narrower than the thorax, with the eyes large, lateral, and densely clothed with very fine hairs; the labial palpi are of moderate length, contiguous and parallel to each other, being obliquely elevated in front of the head, and three-jointed, the middle joint being much the longest, and the third short, and when denuded of its scales and hair, somewhat pointed at the tip. The antennæ are

* More probably Phanessa, being derived from Φάνης, one of the Greek names of Love. Hederich Mythol Lex 1724. Vollm Vollst Wörterb 1816.

rather long, slender, and terminated by an abruptly-formed, short, somewhat cylindrical club, never flattened nor spoon-shaped. The body is very robust, and well-formed for sustaining the powerful flight of these insects. The wings are of large size, with the outer margin not only scalloped, but the anterior have the third and sometimes the last scallop but one, strongly angulated (the tip being, as it were, falcated), and the posterior have the middle of the outer margin also equally angulated. The discoidal cell in both pairs of wings is closed by an oblique vein. The fore legs are very short and rudimentary, so as to be quite unfitted for walking; they are composed of the ordinary parts, except that the tarsal portion is formed into a flat inarticulate plate which, as well as the tibia, is very densely clothed with hairs. The hind feet are long and strong, the tarsi of the ordinary size, five-jointed, and terminated by two curved ungues, on the outside of which is a pair of similarly formed membranous appendages bifid at the base, the under division being very short; between the ungues is a short pulvillus or cushion.*

The caterpillars are long, cylindric, and clothed with numerous bristly spines arranged in whorls round the body, each segment (except that immediately following the head) having a whorl of these spines. The head is generally entire, but in some of the species it is bituberculated. The pupa is considerably angulated, with the head bituberculated, and it is adorned with silvery and golden hues. It is suspended by the tail.

The genus is of considerable extent, but none of the exotic species exceed those of our own country in beauty; indeed, it is impossible to find more exquisite contrasts of colour or delicacy in pencilling than is exhibited by some of our British species. The caterpillars are gregarious in some of the species, but those of the rest live solitarily: the different species of Urtica afford nourishment to the caterpillars of several of them. In their perfect state, several of the species are long-lived, and are often to be seen in the autumn, especially delighting to frequent the dahlia, Michaelmas daisy, and other composite flowers. Ivy also, when in flower, is a particular favourite with them; and some are very fond of ripe fruit, V. Atalanta being even said to be sometimes very destructive to it, especially cherries, by extracting the juice; probably taking advantage of previous injuries occasioned by birds, wasps, and flies. This unusual propensity is occasioned by a very beautiful apparatus forming part of the spiral tongue (or maxillæ), which has recently been described by Mr. G. Newport in his valuable article "Insect," in the Cyclopædia of Anatomy and Physiology. This consists of a great number of minute papillæ along the anterior and lateral margins of the spiral tongue, in the form of little, elongated, barrel-shaped bodies, terminated by three smaller papillæ arranged around their anterior extremity, with a fourth one a little larger than the others, placed in their centre. These papillæ are arranged in two rows along the lateral and anterior surface of each maxilla near its extremity for about one sixth of its whole length, there being seventy-four in each maxilla or half of the spiral tongue. Judging from their structure, and from the circumstance that they are always plunged deeply into any fluid when the insect is taking food, Mr. Newport suggests that they are probably organs of taste. They are largely developed in this genus, but in Pontia and Sphinx Ligustri they are scarcely perceptible. There are also some curious appendages arranged along the inner anterior margin of each maxilla in the shape of minute hooks, which when the proboscis is extended serve to unite the two halves together. In this genus they are described by Mr. Newport as falcated, and furnished with an additional tooth a

* De Geer Mémoires, tom. i. p. 652, and tab. 20. fig. 12, describes and figures the hind tarsi of V. C. Album as furnished with four ungues of equal size and form; and in the Crochard edition of the Règne Animal, Insectes, pl. 135, fig. 3 e, the lateral appendage of the ungues of V. Io, Antiopa, Urticæ, &c. is represented as forming only a simple and undivided piece; but in P. Atalanta these lateral appendages are distinctly bifid, the inner division being about half the length of the exterior.

little beyond the apex, they are so exceedingly minute, and arranged so closely together, that their true form is with difficulty distinguished They lock across each other like the teeth in the jaws of some fishes, and Mr Newport considers that the points of the hooks in one half of the proboscis are inserted when the organ is extended into little depressions between the teeth of the opposite side, so that they form the anterior surface of the canal That they really form the anterior surface of the canal or tube, seems evident from the distinctness with which coloured substances are observed to pass along the tube when the insect is taking food It occasionally happens that some of these insects survive the winter, passing that period of the year in a state of lethargy It has been generally supposed that these are females, which had been produced late in the preceding autumn, and which, although impregnated at that time, had not deposited their eggs, but waited until the renewal of the season brought forth a fresh supply of food for their offspring M Boisduval, however, opposes this, stating that these individuals had entered their lethargic state at a much earlier period (having observed V Polychloros and Urticæ in this state in August), and that their impregnation does not take place until the following spring Mr Brown also opposes the ordinary opinion in Loudon's Magazine of Natural History No 39, founding his observations on the Lepidoptera of Switzerland The Rev W T Bree, however, whose practical knowledge of the subject renders his opinion of so much weight, opposes the statements of Mr Brown, and supports the generally-received opinion in a subsequent number (42) of the same Magazine.

The true Fabrician type of this genus is Pap Io, but Ochsenheimer introduced P Levana and P Cardui (which Fabricius placed in the genus Cynthia) into the genus, forming the latter and P Atalanta into a first section, thus making P Cardui stand as the type of the genus Hubner, also, in his Verzeichniss, gave P Cardui as the true Vanessa, P Atalanta, under the subgeneric name Pyrameis, C Album, under that of Polygonia (since changed by Mr Kirby to Grapta), P Polychloros, Urticæ, and Antiopa under that of Eugonia, P Io, under that of Inachis and P Levana, under that of Araschnia Until, however, a careful revision of all the exotic species belonging to the ill-constructed Fabrician genera Cynthia and Vanessa be made it is impossible to decide on the propriety of the establishment or the extent of these groups

The British species form three evidently natural divisions, which appear to me to be equivalent in value to those which I have proposed amongst the Fritillaries

1 Fore wings with the anal margin very strongly emarginate, posterior wings with a short tail Caterpillars gregarious, with two tubercles on the head C Album

2 Fore wings with the anal margin nearly straight, posterior with a strong angular prominence in the middle of the hind margin Caterpillars gregarious, without the two tubercles on the head Urticæ, Polychloros Antiopa, Io

3 Hind wings rounded and scalloped Caterpillars solitary, without the two tubercles on the head Atalanta

DESCRIPTION OF PLATE XIII

INSECTS —Fig 1 Vanessa C Album (the Comma Butterfly) 2 Showing the under side 3 The Caterpillar 4 The Chrysalis

Fig 5 Vanessa Polychloros (the great tortoise-shell B) 6 Showing the under side 7 The Caterpillar 8 The Chrysalis

Fig 9 Vanessa Urtica (the small tortoise shell B) 10 Showing the under side 11 The Caterpillar 12 The Chrysalis

13 A variety of Vanessa Urticæ

PLANTS —Fig 14 Ribes rubrum (the red currant) 15 Ulmus campestris (the elm) 16 Urtica dioica (the common stinging nettle)

When I first turned my attention to natural history and began to collect insects, I imagined V Polychloros to be nothing more than large faded specimens of V Urtica, and they are certainly very similar at a first glance, but placed side by side, as they will be found in the present plate, the student will have no difficulty in finding sufficient marks of distinction Whilst in the preparatory larva state, the species present but little similarity, which is in fact still less than appears on the plate, for the larva of V. Urticæ, represented with light yellowish markings, becomes nearly all black previous to its change. I have also given the singular variety of V. Urticæ taken near Coventry, and intend, in a supplementary plate at the end of the volume, to give some other varieties of different species which have been communicated to me, and also some new species ascertained to be British since the commencement of this work, particularly Colias Myrmidone, of which Mr Stephens possesses a specimen taken near Dover. H. N. H.

SPECIES 1.—VANESSA C. ALBUM. THE COMMA BUTTERFLY.

Plate xx. fig. 1—4.

SYNONYMES.—*Papilio C. Album*, Linnæus, Lewin Brit. Pap. pl. 9. Donovan Br. Ins. 6, pl. 199. Albin Ins. pl. 54. Harris Aurelian pl. 1. fig. a—d.
Vanessa C. Album, Ochsenheimer, Curtis, Stephens. Duncan Brit. But. pl. 17. fig. 1. Westwood Mod. Class. v. ii. p. 353, fig. 98. 1.
Polygonia C. Album, Hubner (Verz. bek. Schmett.)
Comma C. Album, Rennie Conspect.
Vanessa (s—g. Grapta), Kirby, Fauna Am. Bor. p. 292.

This is the smallest species of the genus, measuring only from 1¾ to rather more than 2 inches in the expanse of its wings. Its form is also quite unlike that of any of the other species, having both the exterior and anal margins of the fore wings strongly emarginate as well as the former scalloped. In its general colour and markings, however, it bears so strong a resemblance to V. Polychloros, that it might at first be easily regarded as a distorted and stunted variety of that species. The wings above are of a tawny-orange colour, with the broad outer margins dark-coloured. There is a black bar running across the middle, and a broader one at the extremity of the discoidal cell, and between the latter of these and the tip of the wings is another abbreviated and more indistinct dark bar. On the posterior part of the disc of the fore wings are also three round black spots, and a dusky patch near the anal angle. The hind wings are dark at the base, with three black discoidal spots and a row of deep crescents in the broad dusky border. On the under side all the wings are of a greyish ashen colour, with very numerous more or less distinct transverse and irregular dark dashes, and a darker brown irregular bar running across the wings, between which and the outer margin are two irregular rows of dull greenish marks, with a small black dot in the middle (these markings vary, however, greatly in intensity in different individuals), in addition to which the disc of the hind wings is ornamented with a white mark like a C.

This species is subject to an extraordinary variation in the form of its wings. In some specimens the incision in the outer margin of the fore wings (extending from the first branch of the median vein to the main branch of the postcostal vein) is so deep that it forms nearly a semicircle, whilst in others it is scarcely more than a sextant, the other indentations being equally varied. Mr Haworth alludes to this, observing "Femina paullo pallidior et subinde minus laciniata" (Lep. Brit. p. 26.) The larva is not gregarious, of a brownish-red colour, the back being reddish in front, with the hinder part white; it is remarkable for having the sides of the head produced above into two conical tubercles, which as well as the spines on the segments of the body are bristly. It feeds on various trees and plants, especially hops, nettles, elm, willow, honeysuckle, &c. The chrysalis is fleshy-coloured or brownish, narrowed in the middle, and spotted with gold. Harris says it remains in this state about fourteen days. There are two broods in the year, the first appearing in June, and the second in August or September. The latter brood are said to be of a paler colour than the summer ones.

This is by no means an uncommon species, being generally distributed. Near London, Hertford, York, Fifeshire, &c., are recorded localities; and the Rev. W. T. Bree informs us that in some years it is not uncommon in many parts of Warwickshire. De Geer (who as well as Reaumur, Memoires, tom. i. pl. 27, has given a very exact account of this species in its different states, Mémoires, tom. i. p. 298, pl. 20, fig. 1—12) observes that it evidently passes the winter in the perfect state, as specimens are occasionally observed in the first days of spring

SPECIES 2.—VANESSA POLYCHLOROS. THE GREAT TORTOISE-SHELL BUTTERFLY.

Plate viii fig 5—8

Synonymes.—*Papilio Polychloros*, Linnæus. Haworth. Lewin Pap. pl. 2. Donovan Brit. Ins. vol. viii. pl. 278. Albin Ins. pl. 55. Wilkes, pl. 108.

Vanessa Polychloros, Ochsenheimer, Curtis, Stephens, Duncan Brit. Butt. pl. 17, fig. 2.

Lugonia Polychloros, Hubner (Verz. bek. Schmett.)

This species is larger than either the preceding or following, with both of which it agrees in the general character of its markings, the wings measuring from $2\frac{1}{2}$ to 3 inches in expanse. On the upper side all the wings are of a dull orange-colour, darker at the base. The anterior have four black subquadrate spots on the posterior part of the disc, and three larger abbreviated fasciæ on the costal edge. The outer margin is dark, with an irregular pale line. The hind wings have a large black costal spot and the outer margin is obscure, with dull blue crescents, and two slender pale lines, parallel to the margin. The under sides of all the wings are clouded with numerous fine black transverse streaks and lines, the basal half being darkest, or rather, there is a very broad ash-coloured fascia beyond the middle of both wings. Beyond this, and parallel with the outer margin, is a row of dull bluish lunules; the hind wings have a small white dot in the middle. There are several varieties, arising from the greater or less extent of the black markings.

The caterpillar feeds on the elm, and is gregarious, at least previous to the first moulting of the skin, the young brood living beneath a common silken web. It is blackish or brownish, with a lateral yellow line, and the spines subramose and yellow. The chrysalis is flesh-coloured, with golden spots, and is attached to the bark of the trees on which the larvæ feed.

The perfect insect appears in the middle of July *, but some individuals survive until the following spring, when they appear in a faded state. It is occasionally very abundant, breeding in the environs of the metropolis where elms abound. I have taken it at Chelsea, and it used to be found in Copenhagen Fields, and numerous other localities in the South of England have been given. Mr Duncan also says that it had been found as far north as Dunkeld, and in many intervening places. It is, however, very uncertain in its appearance. Réaumur has given ample illustrations of the transformations of this species in his Mémoires, tom. i. pl. 23.

SPECIES 3.—VANESSA URTICÆ. THE SMALL TORTOISE-SHELL BUTTERFLY.

Plate xiii fig 9—13

Synonymes.—*Papilio Urticæ*, Linnæus, Lewin Pap. pl. 3. Donovan Brit. Ins. vol. ii. pl. 55. Albin Ins. pl. 4, f. 6. Wilkes Ins. pl. 107. Harris Aurelian, pl. 2, fig. i—n.

Vanessa Urticæ, Fabricius, Ochsenheimer, Stephens. Duncan Brit. Butt. pl. 19, fig. 1.

Lugonia Urticæ, Hubner (Verz. bek. Schmett.)

This very beautiful but most abundant species varies in the expanse of its wings from $1\frac{3}{4}$ to $2\frac{1}{2}$ inches. The wings above are of a rich orange colour; the anterior dark at the base, with three short broad costal bars, between which the ground colour of the wings is paler; behind these are three unequal-sized round spots. The exterior margin of all the wings is black, with a row of blue lunules, and two pale slender parallel submarginal lines. The basal half of the hind wings is also black. Beneath, the orange colour is replaced by pale stone colour and the two smaller posterior discoidal spots are wanting. The margins of all the wings on this side are freckled with brown, having a row of black lunules. Various varieties have been described and figured in which the black

* On the Continent, it is stated to appear in the spring and at the close of the summer; but I apprehend that the early spring specimens are the remnants of the preceding years, and not a distinct brood.

spots are either more or less obliterated, or are enlarged, so as to become confluent A fine individual of the latter kind is figured by the Rev W T Bree in the New Series of the Magazine of Nat Hist Suppl pl 15, and our fig 13, in which the second and third costal black bars are united, whilst the two round discoidal spots are wanting, the hind wings are uniformly obscure

The caterpillars of this species are found on the common nettle in the beginning of June and the middle of August, they are gregarious in the early period of their lives, and are dusky-coloured, varied with green and brown, with paler lines down the back and sides, and with the head black, the body beset with strong branched black spines The chrysalis is brownish, with golden spots on the neck and sometimes entirely golden This golden appearance, which suggested to the early naturalists the names of Chrysalis from the Greek, and Aurelia from the Latin names for gold, and which is so conspicuous in the pupæ of this and the other species of this genus, is owing simply to the shining white membrane immediately below the outer skin, which being of a transparent yellow, gives a golden tinge to the former Its appearance, however, was seized upon by the alchemists as a natural argument in favour of the transmutation of metals, nor was it until the researches of Réaumur in France, and of Ray and Lister in England, that its real nature was discovered, the last-named author having imitated it by putting a small piece of black gall in a strong decoction of nettles, this produces a scum, which, when left on cap-paper, will exquisitely gild it, without the application of the real metal Réaumur also mentions that, for producing this appearance, it is essential that the inner membrane of the chrysalis should be moist, whence may be explained the disappearance of the gilding so soon as the fluids within the body have been absorbed by the formation of the limbs of the butterfly (British Cyclop, art Aurelia)

The perfect insect is very abundant, and appears in the beginning of July and September, often surviving the winter, and coming abroad the first warm days, having been noticed in the Isle of Wight even so early as the 8th of January It is distributed all over the kingdom, extending to the northern extremity of Scotland, in which country it is known under the name of the Devil's or Witch's Butterfly ! In the south of Europe it continues on the wing through the winter, and according to Mr Brown (Mag Nat Hist No 9) it would appear that none of the specimens of this species hybernate in Switzerland and re-appear in the spring

Mr Stephens possesses a most remarkable specimen of this species, having five wings, the fifth of small size, being implanted on the disc of one of the hind wings, which it resembles in its markings It was captured by Mr Doubleday near Epping.

This species afforded the great anatomist Swammerdam materials for a most elaborate memoir on the structure of the larva, and the mode of its transformation to the pupa state His figures occupy two folio plates (34 and 35) in his great work on insects

DESCRIPTION OF PLATE XIV

INSECTS —Fig 1 Vanessa Io (the Peacock Butterfly) 2 Showing the under side 3 The Caterpillar 4 The Chrysalis

„ Fig 5 Vanessa Antiopa (the Camberwell Beauty) 6 The Caterpillar

PLANTS —Fig 8 Salix Russelliana (the Bedford Willow) 9 10 Urtica dioica (the Stinging Nettle)

The little Cynthia Hamsteadiensis, of Petiver, appears out of its place in this plate, but the next (where it would not have appeared much less so) was too crowded to admit it, and I did not like to omit it altogether here particularly as Mr Stephens has inserted it in his work after C Cardui It, however, has so much the air of a species or variety of Hipparchia, that it would have looked much more at home in one of the plates illustrative of that genus I have shown the dingy under side of V Io, as affording a singular contrast to the gay colouring of the upper surface, but that of V Antiopa I considered scarcely worthy of a figure, it is very similar to that of V. Io, with the exception that its pale and dark borders are both there repeated, which renders it less singular H N H

(To be completed in about Sixteen Numbers.)

No. VI.] [Price 2s. 6d.

BRITISH INSECTS AND THEIR TRANSFORMATIONS.

BRITISH BUTTERFLIES

AND

Their Transformations.

ARRANGED AND ILLUSTRATED IN A SERIES OF PLATES

BY H. N. HUMPHREYS, ESQ.

WITH CHARACTERS AND DESCRIPTIONS

BY J. O. WESTWOOD, ESQ., F.L.S.,

SEC. OF THE ENTOMOLOGICAL SOCIETY, ETC. ETC.

LONDON:

WILLIAM SMITH, 113, FLEET STREET.

MDCCCXL.

WORKS PUBLISHED BY WILLIAM SMITH, 113, FLEET STREET.

SMITH'S STANDARD LIBRARY.

In medium 8vo, uniform with Byron's Works, &c

Works already Published

THE LADY OF THE LAKE 1s
THE LAY OF THE LAST MINSTREL 1s
MARMION 1s 2d
THE VISION OF DON RODERICK BALLADS AND LYRICAL PIECES By SIR WALTER SCOTT 1s
THE BOROUGH By the REV G CRABBE 1s 4d
THE MUTINY OF THE BOUNTY 1s 4d
THE POETICAL WORKS OF H KIRKE WHITE 1s
THE POETICAL WORKS OF ROBERT BURNS 2s 6d
PAUL AND VIRGINIA, THE INDIAN COTTAGE, and ELIZABETH 1s 6d
MEMOIRS OF THE LIFE OF COLONEL HUTCHINSON, GOVERNOR OF NOTTINGHAM CASTLE during the Civil War By his Widow, Mrs LUCY HUTCHINSON 2s 6d
THOMSON'S SEASONS, AND CASTLE OF INDOLENCE 1s
LOCKE ON THE REASONABLENESS OF CHRISTIANITY 1s
GOLDSMITH S POEMS AND PLAYS 1s 6d
——— VICAR OF WAKEFIELD 1s
———. CITIZEN OF THE WORLD 2s 6d
———. ESSAYS, &c 2s
*** These four Numbers form the Miscellaneous Works of Oliver Goldsmith, and may be had bound together in One Volume, cloth-lettered, price 8s
KNICKERBOCKER'S HISTORY OF NEW YORK 2s 3d
NATURE AND ART By MRS INCHBALD 10d
A SIMPLE STORY By MRS INCHBALD 2s
ANSON'S VOYAGE ROUND THE WORLD 2s 6d
THE LIFE OF BENVENUTO CELLINI 3s
SCHILLER'S TRAGEDIES THE PICCOLOMINI, and THE DEATH OF WALLENSTEIN 1s 8d
THE POETICAL WORKS OF GRAY & COLLINS 10d
HOME By MISS SEDGWICK 9d
THE LINWOODS By MISS SEDGWICK 2s 6d
THE LIFE AND OPINIONS OF TRISTRAM SHANDY By LAURENCE STERNE 3s
INCIDENTS OF TRAVEL IN EGYPT, ARABIA PETRÆA, AND THE HOLY LAND By J L STEPHENS, Esq 2s 6d
INCIDENTS OF TRAVEL IN THE RUSSIAN AND TURKISH EMPIRES By J L STEPHENS, Esq 2s 6d
BEATTIE'S POETICAL WORKS and BLAIR'S GRAVE 1s
THE LIFE AND ADVENTURES OF PETER WILKINS A CORNISH MAN 2s 4d
UNDINE A MINIATURE ROMANCE Translated from the German, by the REV THOMAS TRACY 9d
THE ILIAD OF HOMER Translated by ALEX POPE 3s
ROBIN HOOD, a Collection of all the ancient Poems, Songs, and Ballads now extant, relative to that celebrated English Outlaw to which are prefixed, Historical Anecdotes of his Life Carefully revised, from RITSON 2s 6d
THE LIVES OF DONNE, WOTTON, HOOKER HERBERT, AND SANDERSON Written by IZAAK WALTON 2s 6d
THE LIFE OF PETRARCH By MRS DOBSON 3s
MILTON'S PARADISE LOST 1s 10d
——— PARADISE REGAINED, and MISCELLANEOUS POEMS 2s
RASSELAS By DR JOHNSON 9d
ESSAYS ON TASTE By the Rev ARCHIBALD ALISON, LL B 2s 6d
THE POETICAL WORKS OF JOHN KEATS 2s
THOMSON'S POEMS AND PLAYS COMPLETE 3s

FOUR VOLUMES ARE NOW ARRANGED ON THE FOLLOWING PLAN —

ONE VOLUME OF "POETRY,"

CONTAINING

SCOTT'S LAY OF THE LAST MINSTREL
SCOTT'S LADY OF THE LAKE
SCOTT'S MARMION
CRABBE'S BOROUGH
THOMSON'S POETICAL WORKS
KIRKE WHITE'S POETICAL WORKS
BURNS'S POETICAL WORKS

A SECOND VOLUME OF "POETRY,"

CONTAINING

MILTON'S POETICAL WORKS
BEATTIE'S POETICAL WORKS
BLAIR'S POETICAL WORKS
GRAY'S POETICAL WORKS
COLLINS'S POETICAL WORKS
KEATS'S POETICAL WORKS
GOLDSMITH'S POETICAL WORKS

ONE VOLUME OF "FICTION,"

CONTAINING

NATURE AND ART By MRS INCHBALD
HOME By Miss SEDGWICK
KNICKERBOCKER'S HISTORY OF NEW YORK By WASHINGTON IRVING
PAUL AND VIRGINIA By ST PIERRE
THE INDIAN COTTAGE By ST PIERRE
ELIZABETH By MADAME COTTIN
THE VICAR OF WAKEFIELD By OLIVER GOLDSMITH
TRISTRAM SHANDY By LAURENCE STERNE

ONE VOLUME OF "VOYAGES AND TRAVELS,"

CONTAINING

ANSON'S VOYAGE ROUND THE WORLD
BLIGH'S MUTINY OF THE BOUNTY
STEPHENS'S TRAVELS IN EGYPT, ARABIA PETRÆA, AND THE HOLY LAND
STEPHENS'S TRAVELS IN GREECE, TURKEY, RUSSIA, AND POLAND

These Volumes may be had separately, very neatly bound in cloth, with the contents lettered on the back, price 10s 6d each

Those parties who have taken in the different works as they were published, and who wish to bind them according to the above arrangement, may be supplied with the title-pages gratis, through their bookseller, and with the cloth covers at 1s each

THE

ADIES' MAGAZINE OF GARDENING.

BY MRS LOUDON

IN MONTHLY NUMBERS ROYAL 8vo, WITH COLOURED PLATES

he First Number will be published on January 1st, 1841, price EIGHTEEN PENCE

As this work is intended solely for Amateurs, and especially for the use of dy Gardeners, I cannot tell exactly of what it will consist till I see what readers desire I have, however, formed the following general plan for nducting it, though, of course, it may be greatly modified and altered by cumstances

I shall give in each Number a Plate, drawn and coloured in the same style those in my LADIES' FLOWER-GARDEN, of one or more Ornamental Plants, uch either are in the country, or may easily be obtained In choosing these, hall studiously avoid all garden varieties of Dahlias and other Florists' owers, and I shall endeavour never to figure any plants that have already peared in any of the English Botanical Periodicals With each Plate ured I shall give a Scientific and Popular Description, with any particulars at I may be able to obtain of it, with regard to its culture and management and the scientific part of this description will be nearly all the science at I shall attempt to introduce into the work I shall then give any Papers at may be sent to me on Floriculture, on Laying Out or Decorating Flower-ardens, or on any other subjects that my correspondents may choose to write on and I shall conclude the First Part with two Papers—one on the History d Culture of some Favourite Plant which may be in flower at that time, d the other, on some Insect or other Animal connected with gardens

The Second Part will consist of Reviews Queries and Answers, Retrospective Criticism, Visits to the Nurseries, Extracts from Gardening Books, c, concluding with a Floral Calendar of Plants in Flower, and Operations Culture, for the current month

Having thus given a rapid sketch of my plan, I have only to invite all Ladies nd of Gardening to send me the results of their own experience, or any uestions as to Floriculture, Laying Out or Planting Flower-Gardens, Names Plants or Insects, &c, which they may wish answered, and I think I may nture to promise that I shall generally be able to obtain for them the information they desire

I have only to add, that all Communications, Books, &c, intended for the LADIES' MAGAZINE OF GARDENING, must be addressed to me, at my Publisher's, 13, Fleet-street

J W LOUDON

Bayswater, November 20th, 1810

LONDON BRADBURY AND EVANS, PRINTERS, WHITEFRIARS

SPECIES 4—VANESSA ANTIOPA THE WHITE BORDER, OR CAMBERWELL BEAUTY

Plate xiv fig 5—6

SYNONYMES —*Papilio Antiopa*, Linnæus Haworth, Lewin Papil pl 1 Donovan Brit Ins vol iii pl 89 Harris Aurel pl 12, fig a—e Wilkes, pl 113 | *Vanessa Antiopa*, Ochsenheimer, Stephens Duncan Brit Butt pl 18, fig 2 Curtis Brit Ent vol ii pl 96 (V Antiope) *Eugonia Antiopa*, Hubner (Verz bek Schmett)

This fine species varies in the expanse of its wings from 2_6 to $3\frac{1}{2}$ inches The wings are on the upper side of a rich claret black, with the apical margin and two costal spots near the extremity of the fore wings, of a white or whitish colour, slightly speckled with black, the white margin is preceded by a series of blue spots, on a black bar Beneath, the wings are dark brown, with a very great number of slender transverse black lines The white margin and costal spots are as on the upper side, but the black subapical bar, with its blue spots, is almost obliterated The hind wings are marked in the centre with a minute white spot The pale margin of the wings varies to deepish yellow The caterpillar, which is gregarious, is of a black colour with squarish dorsal spots, and the abdominal prolegs of a red colour * It feeds on the willow and birch, always selecting the highest branches, according to Harris and is found at the beginning of July The chrysalis is blackish, spotted with fulvous, and is dentated The perfect insect appears at the beginning of August, but sometimes survives the winter, and deposits its eggs in the following spring It appears to be distributed nearly over the whole of the kingdom having been found as far north as Ayrshire It must now, however be considered as one of our rare butterflies, although about seventy years ago it appeared in such immense numbers throughout the kingdom that the Aurelians of that day thence gave it the name of the Grand Surprise Since that period, however, it has become rare, appearing, however, periodically, after a lapse of eight, ten, or more years " To suppose they come from the Continent is an idle conjecture, because the English specimens are easily distinguished from all others by the superior whiteness of their borders Perhaps their eggs in this climate, like the seeds of some vegetables, may occasionally lie dormant for several seasons, and not hatch until some extraordinary but undiscovered coincidences awake them into active life " (Haworth Lep Britann p 28) It received its English name of the Camberwell Beauty from having been observed at that village, to which it was attracted by the willows, which grew there in profusion

---

SPECIES 5—VANESSA IO THE PEACOCK BUTTERFLY

Plate xiv fig 1—4

SYNONYMES —*Papilio Io* Linnæus, Haworth, Lewin Papil pl 4 Donovan Brit Ins pl 206 Albin pl 4, f 5 Wilkes, pl 106 Harris Aurelian, pl 8, fig f—k | *Vanessa Io*, Fabricius, Ochsenheimer, Stephens, Duncan Brit Butt pl 18, fig 1 *Inachis Io*, Hubner (Verz bek Schmett)

This very beautiful insect, which measures from $2\frac{1}{2}$ to 3 inches in the expansion of its wings, may be considered as one of the commonest of our butterflies The fore wings on the upper side are of a dark but rich red colour The costa is varied with black and yellowish buff patches, the base of the costa being marked with black and yellowish transverse streaks. Near the apex of the wings is a very large eye, in which red, black, yellowish-buff, and leaden blue are agreeably blended The outer margin of the wing is dark brown, and there are five blue spots, three of which appear in the eye and two below it The hind wings are of a darker red, the

* De Geer has illustrated the transformation of this species in his Mémoires, tom i pl 21, and has figured several varieties in the spines of the larvæ, these spines do not exist on the segment succeeding the head

base and apex being brown; near the outer angle is a very large eye, with a black centre, in which are several blue markings. This is surrounded by a whitish circle, which is deeply margined with black towards the base of the wing. All the wings beneath are dark brown, with black transverse streaks; the anterior having five small pale marks, representing the blue dots of the upper side, and the posterior having a broad central darker bar, margined with black, within which is a small central white spot.

The caterpillar, which is gregarious, spinose, black, spotted with white, and with the hind legs red, feeds on the common stinging-nettle, and is found at the beginning of July. The chrysalis is greenish, dotted with gold, and dentated. The imago appears in the middle of July, and often survives until the following spring, when the female deposits its eggs. Although very abundant in England, it appears not to extend further north than the Firth of Forth; and in the south of Scotland it is but sparingly seen.

This butterfly and its preparatory states have formed the subject of one of the most interesting of the "Memoires" of Reaumur, by whom it was selected as an example to illustrate the manner in which the butterflies which are merely suspended by the tail in the chrysalis state effect their transformations. If the proceedings of the swallow-tail or cabbage butterflies on assuming the pupa state (see *ante*, pp. 9 and 23) have excited our admiration, the mode in which these caterpillars change to *suspended* chrysalides is far more extraordinary. Like the former, each constructs a small button of silk, to which it firmly attaches itself by the hooks of the hind feet. When this is effected, the head is permitted to hang downwards. Whilst thus suspended, it succeeds, after at least twenty-four hours' contortion, in forming a slit down its back, through which the head of the chrysalis is protruded, and the caterpillar skin gradually pushed upwards to the tail. A delicate operation has still to be performed: the caterpillar was suspended by the hooks of its own hind legs to the silken button; but not only has the still partially enclosed chrysalis to disengage itself entirely from the skin of the caterpillar, and *attach itself* to the silken button, but also to get rid of the old and no longer necessary caterpillar skin. To effect these objects, the chrysalis carefully withdraws its tail from the skin, seizing hold of the outside of the latter by pressing two of the rings of its body together, and enclosing between them part of the old skin. By repeating this proceeding, it at length pushes its tail upwards, till it reaches the silken button, to which it fastens itself by means of the hooks with which the tail of the chrysalis is furnished. We now see the chrysalis suspended head downwards, by the side of the old caterpillar skin, which it ultimately gets rid of by a succession of gyrations, which burst the silken threads holding the caterpillar skin, and which, no longer supported, falls to the ground.

DESCRIPTION OF PLATE XV.

INSECTS.—Fig. 1. Vanessa Atalanta (the Red Admiral Butterfly). 2. Showing the under side. 3. The Caterpillar. 4. The Chrysalis.

Fig. 5. Cynthia Hunteri. 6. Showing the under side.

" Fig. 7. Cynthia Cardui. 8. Showing the under side. 9. The Caterpillar. 10. The Chrysalis.

PLANTS.—Fig. 1, 2. Urtica dioica (the Stinging Nettle). 3. Cnicus lanceolatus (the Spear-plume Thistle).

Cynthia Hunteri has no greater claim to be considered a British species than Argynnis Aphrodite. But, as in that case, I avail myself of the fact of its having been taken in England, to introduce so beautiful an insect in this work. The under side presents an extremely elegant variation of the colouring of C. Cardui. I am compelled unavoidably to give here a third portrait of the uncomely nettle, as it is the only food of the insects which it accompanies.—H. N. H.

SPECIES 6—VANESSA ATALANTA THE RED ADMIRAL, OR ALDERMAN BUTTERFLY

Plate xv fig 1—4

SYNONYMES —*Papilio Atalanta*, Linnæus Haworth, Lewin P pl 1 pl 7 Donovan vol viii pl 260 Alb , pl 3 Wilkes, pl 105 Harris Aurelian, pl 6 f a—h | *Vanessa Atalanta* Fabricius, Stephens, Curtis Duncan Brit Butt pl 20, fig 1 | *Pyrameis Atalanta*, Hubner (Verz bek Schmett) *Ammiralis Atalanta* Rennie

This remarkably rich-coloured butterfly is one of the commonest of our native species It varies in the expanse of its wings from 2½ to 3 inches The ground colour of the upper surface of the fore wings is intense velvety blue-black, brownish at the base, having an irregular oblique central bar of bright red, slightly curved on the side nearest the tip of the wing, and formed as it were of large squarish confluent patches, it does not quite extend to the anal angle of these wings Between the fascia and the apex of the fore wings is a large costal white spot beyond which is a curved row of five white spots, of which the first and fourth are the largest Still nearer the margin of the wing is an obscure bluish wave The hind wings are blackish brown above, with a broadish red margin, in which are four black dots, and there are two obscure confluent blue spots at the anal angle On the under side the fore wings are black, the base with several narrow red and bluish transverse stripes, the red oblique bar is here present but more broken, between which and the large costal white spot is a horse-shoe blue mark The apex is ashy-brown, with two small brown eyes with white centres and two white spots The hind wings on this side are brown and most beautifully mottled with black and grey, with a large triangular pale spot in the middle of the costal margin, and two transverse and wedge-shaped discoidal black marks Near the margin of the wing is a row of four obscure eye-like patches In some specimens the red bar of the fore wings bears a small white dot near its hinder extremity these, according to Mr Haworth, are the females

This species differs from all the foregoing, not only in the form of the wings of which the anterior are less strongly angulated, and the posterior rounded, but also in several other characters, especially the form of the palpi and the habit of the caterpillars Hence Mr Kirby suggests in the Fauna Boreali Americana (p 294) that it "seems rather to belong to the genus, or perhaps sub-genus, Cynthia, at any rate it forms a connecting link between it and Vanessa"

The caterpillar is of a dusky-green colour, with a yellowish dorsal line and also a pale line on each side above the feet The chrysalis is brownish or blackish, beneath grey with golden spots

The caterpillar feeds on the common nettle, especially preferring the seeds, and is found in July the imago is abundant wherever this plant is common—it appears at the beginning of August, and survives the winter, the female depositing her eggs in the following spring

According to Sepp, the caterpillar shortly after it is hatched selects a nettle-leaf, which it draws together with threads into a roundish hollow form, leaving for the most part an opening into the interior both before and behind, thus serving both for shelter and food until it is almost devoured, when it selects a fresh leaf and proceeds with it in the same manner, one caterpillar only being found on a single leaf thus indicating a peculiar liking for a solitary life, a circumstance confirmed by the eggs being laid singly and apart, whereas caterpillars hatched from eggs deposited in clusters are gregarious The caterpillar state lasts about five weeks

The species appears to be very widely distributed I have received specimens from North America, which, although slightly differing from our native individuals, I cannot regard as specifically distinct Such is also the opinion of Mr Kirby, who has described his American specimens under this name

It also occurs throughout Europe and along the African shores of the Mediterranean It delights in the flowers of the ivy and dahlia, and is a remarkably bold insect, whereof some remarkable instances are mentioned in Loudon's Magazine of Natural History (No 25)

GENUS XII

CYNTHIA, Fabricius

This genus, or perhaps rather sub-genus, differs chiefly from Vanessa in the form of the wings, the anterior pair being very slightly angulated at the tip, whilst the hind ones are rounded and scalloped, and in certain trivial distinctions, as in the club of the antennæ, which is very short and compressed, and in the palpi, which are long deflexed, pointed and beak-like, the second joint, with the posterior half pilose The caterpillar and chrysalis resemble those of Vanessa By Curtis, it is united with the last-named genus As, however, C Cardui is not one of the types of the genus as established by Fabricius, it is perhaps best to retain it, considering the exotic species Papilio Arsinoe and Œnone as the types of the two sections into which it is divided, and regarding Cardui as an aberrant species leading to Vanessa

SPECIES 1 —CYNTHIA CARDUI THE PAINTED LADY

Plate xv fig 7—10

Synonyms —*Papilio Cardui*, Linnæus, Fabricius, Haworth, Lewin Pap t 6, f 1 4 Donovan's Ins v 9, tab 292 Shaw Nat Miscell 9, tab 430 Panzer Faun Ins Germ 22 19 Wilkes Papil t 107, f 1 Albin Ins t 56 Harris Aurelian t 11, fig c—f *Libythea Cardui*, Lamarck

Vanessa Cardui, Godart, Latreille, Meigen, Hubner (Verz bek Schmett) *Cynthia Cardui*, Fabricius, Kirby (F B A) Stephens, Duncan Brit Butt t 19 f 2

This elegant insect in its markings might at first sight be mistaken for a mottled and faded Atalanta, so closely allied are the two species together, although perfectly distinct both in habits and markings, being in fact widely separated in the Linnæan system, one belonging to the Nymphales Phalerati, and the other (C Cardui) to the N gemmati, in consequence of the wings being marked with eye-like spots It varies in the expanse of its wings from $2\frac{1}{4}$ to $2\frac{3}{4}$ inches The fore wings on the upper side are at the base brown, the disk tawny orange *, with three somewhat square black spots, the apex blackish, with five white spots, the largest of which is on the costa and the four others form a curved line, between which and the margin is a slender whitish line The hind wings above have the base and costal margin brown, the disk fulvous, with numerous black marks arranged, as it were, in four transverse rows, the second forming a row of round darker-coloured spots, the fourth being marginal, the margin itself whitish Beneath, the fore wings are nearly marked as above, but the fulvous colour is more diffused, the dark spots are smaller, and the apex of the wing is dark stone-colour, instead of black The hind wings below are beautifully mottled with pale olive brown, yellowish buff, and white, the veins being white near the hind margin is a row of slender blackish-blue marks, above which are four beautiful eyes, the two middle ones being smaller than the outer ones, which are circled with black The markings vary in size in different individuals, Mr Stephens having described several varieties The caterpillar is spined, of a brown colour, with interrupted lateral yellow lines, it is solitary, and feeds on the Carduus lanceolatus, and other species of the same genus, as well as on the nettle, mallow, artichoke, &c It is found in the middle of July Like that of V Atalanta, it draws up the leaves upon which it is feeding with its threads, and like it is solitary in its habits The chrysalis is brown, with ash-coloured lines and golden spots

* The tawny orange marks on the right forewing bear a tolerably good resemblance to a map of England and Ireland

This is one of those species of butterflies remarkable for the irregularity in its appearance, in some years occurring plentifully even in the neighbourhood of London, after which it will disappear for several years. Indeed, instances are on record in which, owing to the vast numbers, migration has become necessary and in the Annales des Sciences Naturelles for 1828, an account is given of an extraordinary swarm which was observed in the preceding May in one of the cantons of Switzerland, the number of which was so prodigious that they occupied several hours in passing over the place where they were observed The precise causes for this phenomenon were not investigated, and the time of the year is remarkable Like V Atalanta, the species is very widely dispersed, being an inhabitant of North America New South Wales, Java, both extremities of Africa, Brazil, &c

There are numerous notices relative to this butterfly contained in Loudon's Magazine of Natural History Nos 4, 13, 18, 26 31, and 39 to which I must refer the reader

---

SPECIES 2—CYNTHIA HUNTERA HUNTER'S CYNTHIA

Plate xv fig 5, 6

SYNONYMES—*Papilio Huntera* Fabricius Herbst, Abbot and Smith, Ins Georgia, vol i t 9
Papilio Cardui Virginiensis, Drury Ins. i p 15, fig 1
Papilio Iole, Cramer
Cynthia Huntera Kirby Fauna Bor Amer p 296 (1837) Westwood in Drury Ins 2nd edit (1837)
Vanessa Huntera, Dale in Loudon's Mag Nat Hist vol iii p 332 Stephens

This American species, although closely allied to the preceding, cannot be considered as its Transatlantic representative, as Drury imagined by calling it Cardui Virginiensis It measures about 2¼ inches in the expanse of the wings, which are of a less brilliant tawny orange colour than those of P Cardui, brown at the base, the orange disk much broken in the fore wings by blackish irregular bars, the apex blackish with a long white costal spot, and four dots near the apex, white, between which and the margin is a pale broken rivulet Beyond the middle of the hind wings is a slender interrupted brown bar succeeded by four indistinct ocelli, a black submarginal bar, and two very slender marginal dark lines But the great beauty of the insect consists in the under side of the wings, the anterior being elegantly varied with white orange, brown and black, with two eyes near the apex The disk of the hind wings is white, with the veins and many lines and bars of brown, these form a double scallop beyond the middle of the wing, succeeded by a white bar of the same form the terminal part of the wing being brown and ornamented by two very large eyes, margined with black, between these and the margin is a slender bar, and two dark thin marginal lines

The caterpillar of this butterfly is described by Drury as green, with black rings round the body, and as feeding, about New York, on the wild balsam, appearing about the end of July or beginning of August According to Abbot, however, the caterpillar is brown, with the incisions and lateral line yellow it has also two dorsal lines formed alternately of white and red points It feeds upon the Gnaphalium obtusifolium The chrysalis state is assumed at the end of April or beginning of May The butterfly appears in about ten days Like C Cardui, its caterpillar folds and spins the leaves of the plants on which it feeds, together, and the perfect insect appears about once in five or six years in very great abundance, at other times they are scarce

An instance of the capture of this butterfly by the late Captain Blomer in Pembrokeshire has been recorded by Mr Dale in the work above mentioned

CYNTHIA HAMPSTEDIENSIS. ALBIN'S HAMPSTEAD-EYE.

Plate xiv, fig. 7.

SYNONYMS.—*Papilio oculatus Hampstediensis ex aureo fuscus*, Petiver Pap. pl. 5. f. 2. Haworth, p. 34. | *Hipparchia Hampsteadiensis*, Jermyn. *Cynthia Hampstediensis*, Stephens Ill. pl. 5. f. 3, 4. Brit. Batt. 20.

This butterfly is represented by Petiver, the only authority for the species, as about 2 inches in expanse, or of the size of Hipparchia Ægeria: the fore wings are brown with three transverse sub-costal spots, two elongate ones near the hinder margin, and the margin itself yellow; at the apical and anal angle is a large ocellus. The hind wings are brown, with a yellow margin, and with two large ocelli near the hind margin. Beneath the fore wings are yellowish, with brown cloudings, and with a row of brown sub-marginal lunules. The posterior are dull yellow, with darker cloudings of brown at the base, with a small ocellus near the anal angle, and a row of four brown spots, between which and the margin is a nearly obsolete row of brown lunules.

The only instance of the capture of this otherwise unknown insect was recorded more than a century ago by the faithful Petiver, from whose representation our figure is taken; its capture is noticed by him in these words: "Albin's Hampstead Eye, where it was caught by this curious person, and is the only one I have yet seen." Like Aphrodite and Huntera, I have no doubt that it is an exotic species which had been accidentally brought to this country. I have followed Mr. Stephens in giving this as a species of Cynthia rather than as an Hipparchia, as it is evidently allied to Cynthia Orythia.

CYNTHIA LEVANA (Papilio L., Linn., Vanessa L., Ochsenh.), a Continental species, being 1½ inches in the expanse of its wings (which are fulvous-coloured, varied with black and yellow above, the anterior having also several white spots; and beneath reticulated with whitish yellow, fulvous, brown and yellowish,) is indicated as British by Turton, Syst. Nat. p. 42, but no specimen captured in this country is known to be in existence.

GENUS XIII.

APATURA*, FABRICIUS.

DESCRIPTION OF PLATE XVI.

INSECTS.—Fig. 1. Apatura Iris (the Purple Emperor), male. 2. The female. 3. Showing the under side. 4. The Caterpillar. 5. The Chrysalis.

" Fig. 6. Limenitis Camilla (the White Admiral). 7. Showing the under side. 8. The Caterpillar. 9. The Chrysalis.

PLANTS.—Fig. 10. Quercus Sessiliflora (the Sessile-fruited Oak).

" Fig. 11. Lonicera Periclymenum (the Common Honeysuckle).

The Purple Emperor is the most favourite butterfly of English collectors, partly from its comparative rarity and the difficulty of capture even when discovered, and partly in consequence of its being the only British Butterfly of large size that exhibits a blue tint. The beautiful purple gloss exists, however, only in the male, and only in certain lights. I have endeavoured to have the beautiful effect it produces imitated in the fig. 1, of Plate 16, but art cannot do justice to the fitful flashes of rich colour which every change of light produces upon this insect, leaving it at one moment its all sober brown, the colour of the female, and the next tinging it with a flush of the richest metallic purple. As in flying downward the purple is nearly constant over a large portion of the wings, I have chosen that position as the one most capable of conveying, in a drawing, some idea of the beauty of the insect. The female, generally somewhat larger in size, and of a paler brown, is represented beneath

---

* More properly Apatura, a name of Venus, from απατη. Voller Vollst. Worterb. p. 271.

Only this solitary species of the genus Apatura is found in England but on the Continent I have met with several others particularly the beautiful 'Mars Changeant' of the south of France, which, though of a brighter brown, greatly resembles the present species, both in the purple gloss, and in the markings of the under side Limenitis Camilla has been placed in this Plate to show the affinity which exists between these two genera, on comparing the under sides The Butterflies are from fine specimens in the British Museum The Caterpillars and Chrysalides are from Hubner and Godart H N H

The insects composing this splendid genus are at once distinguished from all the preceding genera of this family by having the antennæ very gradually thickened towards the tips into a club, whilst it is separated from Hipparchia by their being straight and not curved, and by the robust structure of the insects It agrees with the Fritillaries in having naked eyes, by which character it is at once separated from Limenitis The palpi are close together and compressed, so as to form an elongated beak pointed at the tip The body is robust, the wings powerful, the anterior having the posterior margin entire, and the hind wings scalloped The discoidal cell of the wings is not closed, the fore legs are rudimental, with the tarsi articulated thus differing from Vanessa The four hind legs are terminated by two strong ungues, defended at the side by bifid membranous appendages

The larva somewhat resembles a slug, having the body thickest in the middle, fleshy, destitute of spines except a pair on the crown of the head and the bifid tail The chrysalis is compressed, with the head-case bifid This variation in the form of the larva has induced Dr Horsfield to unite this genus with Hipparchia and some others into a distinct primary division of the diurnal Lepidoptera named Thysanuromorpha, from a supposed resemblance to the forked-tailed Thysanuræ, or spring-tailed insects! The only British species is the following

SPECIES 1 —APATURA IRIS THE PURPLE EMPEROR

Plate xvi fig 1—5

SYNONYMES —*Papilio Iris*, Linnæus, Haworth Donovan pl 37 Lewin Papil pl 16 Wilkes, pl 120 Harris Aurelian pl 3, fig sup

Apatura Iris, Ochsenheimer, Leach, Stephens Curtis, pl 338 Duncan, Brit Butt pl 21

Doxocopa Iris, Hubner (Verz bek Schmett)

This fine insect varies in the expanse of its wings from $2\frac{1}{2}$ to $3\frac{1}{4}$ inches The wings of the male are above of a blackish hue, with a splendid purple blush varying according to the position from which they are seen, and marked in the middle and towards the hinder margin with white spots, the inner ones forming the curved upper extremity of a bar which runs across the hind wings nearly to the anal angle, this angle itself being orange, with two black spots, above which is an ocellus The under side of the fore wings is varied with grey, orange, fulvous, and black, there being an interrupted, curved white fascia across the wings, behind which is a black eye with a lilac centre surrounded by a broad orange circle in which are two white spots The hind wings on this side are grey, with a broad white bar attenuated towards the anal angle, on each side broadly ferruginous, the anal angle ferruginous, above which is a black eyelet with a lilac pupil and orange iris

The wings of the female are brown, destitute of the purple lustre, but marked as in the male

The caterpillar is green, with pale yellow lateral oblique stripes It feeds on the broad-leaved sallow, and is found at the end of May The chrysalis is of a pale green colour The perfect insect is found in the middle of July in woods, in various parts of the South of England Epping Forest, Great and Little Stour Woods, Wrabness, and Ramsay, Essex, Badly, Dodnash, and Raydon Woods, in Suffolk, Clapham Park Wood, Beds, Bringsop Copse, Heref, Enborne Copse, Berks, near Warminster, Wilts, Christchurch, Hants, Monkswood, Camb, near

Hertford, and Coombe and Darenth Woods,—have been given as its localities to which we may add that it is "occasionally though rarely seen in Warwickshire, near Doncaster, and in the Isle of Wight (Rev W T Bree MSS)

Owing to the habit which the Purple Emperor exhibits of fixing his throne on the summit of a lofty oak from the utmost sprigs of which, on sunny days, he performs his aerial excursions, defending his territory against a rival emperor with the greatest energy, it is necessary to use a bag net fixed at the end of a slender rod twenty or thirty feet long He is exceedingly bold, and will almost suffer himself to be pushed off his seat The females are much rarer, and do not take such lofty flights as the males

---

GENUS XIV
LIMENITIS *, FABRICIUS

This genus is closely allied to Apatura, but differs in its general weaker formation, and in the hinder margin of the fore wings being rounded, and not concave as in the Purple Emperor, the hind wings are more rounded, and the eyes are pubescent By these characters, and by the gradual formation of the straight club of the antennæ, it is distinguished from all the other genera of this family, the palpi are not contiguous as long as the head, not pointed at the tip and clothed with scales and hair, the hind wings have the discoidal cell open, the fore legs are short in both sexes, the tarsi formed of a single joint clothed with long hairs and terminated by a small single unguis the four hind legs are formed as in Apatura The larvæ are long, cylindric, with several pairs of obtuse hirsute spines on the back, and lateral fascicles of hairs The chrysalis has the head also beaked, and is very gibbose beneath It is suspended by the tail The close relation of this genus and Apatura in the perfect state is sufficient to prove that they are not referable to separate primary groups of the Diurnal Lepidoptera, on account of the differences in their caterpillar state

---

SPECIES 1 —LIMENITIS CAMILLA THE WHITE ADMIRAL

Plate xvi fig 6—9

Papilio Camilla, Linnæus, Haworth, Lewin Papil pl 8 Donovan Ins 8, pl 241 Harris Aurelian, pl 30, fig m, n
Papilio Sibilla, Fabricius, Stewart
Limenitis Camilla Leach Curtis Brit Ent pl 124 Duncan Brit Butt pl 20 fig 2 Hubner (Verz bek Schmett)

The wings of this species measure from 2 to 2½ inches in expanse The upper surface is dull black, with a curved interrupted row of white spots extending from near the middle of the costa of the fore wings to the anal angle of the hind ones, in addition to which the anterior have several additional small spots near the apex, and the posterior have an obscure reddish spot at the anal angle, within which are two black dots Beneath the ground colour of the wings is yellowish brick-red, with the white spots of the upper side conspicuous in addition to which all the wings, especially at the base, are marked with black streaks and dots, and the hind wings, between the white band and the margin, have two rows of black dots and two rows of crescents on the margin.

---

* One of the names of Venus Vollm Vollst Wort 1143

ILLUSTRATED EDITION OF FROISSART.

In two thick Volumes, price 36s

SIR JOHN FROISSART'S CHRONICLES OF ENGLAND, FRANCE, SPAIN, &c.

THIS Edition is printed from the Translation of the late THOMAS JOHNES, Esq, and collated throughout with that of LORD BERNERS, numerous additional Notes are given, and the whole embellished with *One Hundred and Twenty Engravings on Wood,* illustrating the Costume and Manners of the period chiefly taken from the illuminated MS copies of the Author, in the British Museum, and elsewhere

COMPANION TO FROISSART.

In two Volumes, price 30s

THE CHRONICLES OF MONSTRELET.

WITH NOTES AND WOODCUTS, UNIFORM WITH THE ABOVE EDITION OF FROISSART

ILLUSTRATED EDITION OF "MARMION"

In demy 8vo, price 16s cloth, 21s morocco elegant,

MARMION.

A Poem

BY SIR WALTER SCOTT

WITH FIFTY BEAUTIFUL WOOD-ENGRAVINGS

ILLUSTRATED WITH FIFTY-ONE PORTRAITS,

BURNET'S HISTORY OF HIS OWN TIMES.

In Two Volumes, super-royal 8vo, cloth lettered, price 2*l* 2s, or half bound in morocco, 2*l* 12s 6d.

ILLUSTRATED WITH FIFTY-SIX PORTRAITS,

CLARENDON'S HISTORY OF THE REBELLION,

In Two Volumes, imperial 8vo, price 2*l* 10s cloth lettered

In one Volume, foolscap 8vo, price 7s in cloth,

A TREATISE ON THE INSECTS INJURIOUS TO THE GARDENER, FORESTER, AND FARMER.

TRANSLATED FROM THE GERMAN OF M KOLLAR, AND ILLUSTRATED WITH ENGRAVINGS

BY J AND M LOUDON

WITH NOTES BY J O WESTWOOD, ESQ

"We heartily recommend this treatise to the attention of every one who possesses a garden, or other ground, as we are confident that no one taking an interest in rural affairs can read it without reaping both pleasure and profit from its perusal"—*Literary Gazette*

"We have always wondered that, in a country like this, where the pursuits of agriculture and horticulture are so universal and important, entomologists should never have bethought them of writing a book of this description It is, therefore, with great satisfaction, that we announce the appearance of the present translation of a work which goes far to supply the deficiency we have spoken of"—*Athenæum*

"From the very neat and cheap manner in which the volume is got up, we trust it will become a favourite, not only with the entomologist, but with every lover of agriculture, arboriculture, and horticulture"—*Mag Nat Hist*, Feb 1840

(To be completed in about Sixteen Numbers.)

No. VII.] [PRICE 2s. 6d.

BRITISH INSECTS AND THEIR TRANSFORMATIONS.

BRITISH BUTTERFLIES

AND

Their Transformations.

ARRANGED AND ILLUSTRATED IN A SERIES OF PLATES

BY H. N. HUMPHREYS, ESQ.

WITH CHARACTERS AND DESCRIPTIONS

BY J. O. WESTWOOD, ESQ., F.L.S.,

SEC. OF THE ENTOMOLOGICAL SOCIETY, ETC. ETC.

LONDON:
WILLIAM SMITH, 113, FLEET STREET.

MDCCCXLI.

WORKS PUBLISHED BY WILLIAM SMITH, 113, FLEET STREET.

SMITH'S STANDARD LIBRARY.

In medium 8vo, uniform with Byron's Works, &c

Works already Published

ROKEBY BY SIR WALTER SCOTT. 1s 2d

THE LADY OF THE LAKE 1s
THE LAY OF THE LAST MINSTREL 1s
MARMION 1s 2d
THE VISION OF DON RODERICK, BALLADS AND LYRICAL PIECES By Sir Walter Scott 1s
THE BOROUGH By the Rev G Crabbe 1s 4d
THE MUTINY OF THE BOUNTY 1s 4d
THE POETICAL WORKS OF H KIRKE WHITE 1s
THE POETICAL WORKS OF ROBERT BURNS 2s 6d
PAUL AND VIRGINIA, THE INDIAN COTTAGE and ELIZABETH 1s 6d
MEMOIRS OF THE LIFE OF COLONEL HUTCHINSON, Governor of Nottingham Castle during the Civil War By his Widow, Mrs Lucy Hutchinson 2s 6d
THOMSON'S SEASONS, AND CASTLE OF INDOLENCE 1s
LOCKE ON THE REASONABLENESS OF CHRISTIANITY 1s
GOLDSMITH S POEMS AND PLAYS 1s 6d
——————— VICAR OF WAKEFIELD 1s
——————— CITIZEN OF THE WORLD 2s 6d
——————— ESSAYS, &c 2s

*** These four Numbers form the Miscellaneous Works of Oliver Goldsmith, and may be had bound together in One Volume, cloth-lettered, price 8s

KNICKERBOCKER'S HISTORY OF NEW YORK 2s 3d
NATURE AND ART By Mrs Inchbald 10d
A SIMPLE STORY By Mrs Inchbald 2s
ANSON S VOYAGE ROUND THE WORLD 2s 6d
THE LIFE OF BENVENUTO CELLINI 3s
SCHILLER'S TRAGEDIES, THE PICCOLOMINI and THE DEATH OF WALLENSTEIN 1s 8d
THE POETICAL WORKS OF GRAY & COLLINS 10d
HOME By Miss Sedgwick 9d
THE LINWOODS By Miss Sedgwick 2s 8d
THE LIFE AND OPINIONS OF TRISTRAM SHANDY By Laurence Sterne 3s
INCIDENTS OF TRAVEL IN EGYPT, ARABIA PETRÆA, AND THE HOLY LAND By J L Stephens, Esq 2s 6d
INCIDENTS OF TRAVEL IN THE RUSSIAN AND TURKISH EMPIRES By J L Stephens, Esq 2s 6d
BEATTIE'S POETICAL WORKS, and BLAIR S GRAVE 1s
THE LIFE AND ADVENTURES OF PETER WILKINS, a Cornish Man 2s 4d
UNDINE a Miniature Romance Translated from the German, by the Rev Thomas Tracy 9d
THE ILIAD OF HOMER Translated by Alex Pope 3s
ROBIN HOOD, a Collection of all the ancient Poems, Songs, and Ballads now extant, relative to that celebrated English Outlaw, to which are prefixed, Historical Anecdotes of his Life Carefully revised, from Ritson 2s 6d
THE LIVES OF DONNE, WOTTON, HOOKER, HERBERT, AND SANDERSON Written by Izaak Walton 2s 6d
THE LIFE OF PETRARCH By Mrs Dobson 3s
MILTON'S PARADISE LOST 1s 10d
——————— PARADISE REGAINED, and MISCELLANEOUS POEMS 2s
RASSELAS By Dr Johnson 9d
ESSAYS ON TASTE By the Rev Archibald Alison, LL B 2s 6d
THE POETICAL WORKS OF JOHN KEATS 2s
THOMSON S POEMS AND PLAYS, Complete 5s

FOUR VOLUMES ARE NOW ARRANGED ON THE FOLLOWING PLAN —

ONE VOLUME OF "POETRY,"

CONTAINING

SCOTT'S LAY OF THE LAST MINSTREL
SCOTT'S LADY OF THE LAKE
SCOTT'S MARMION
CRABBE'S BOROUGH
THOMSON S POETICAL WORKS
KIRKE WHITE'S POETICAL WORKS
BURNS S POETICAL WORKS

A SECOND VOLUME OF "POETRY,"

CONTAINING

MILTON S POETICAL WORKS
BEATTIE'S POETICAL WORKS
BLAIR S POETICAL WORKS
GRAY S POETICAL WORKS
COLLINS S POETICAL WORKS
KEATS'S POETICAL WORKS
GOLDSMITH S POETICAL WORKS

ONE VOLUME OF "FICTION,"

CONTAINING

NATURE AND ART By Mrs INCHBALD
HOME By MISS SEDGWICK
KNICKERBOCKER'S HISTORY OF NEW YORK By WASHINGTON IRVING
PAUL AND VIRGINIA By ST PIERRE
THE INDIAN COTTAGE By ST PIERRE
ELIZABETH By MADAME COTTIN
THE VICAR OF WAKEFIELD By OLIVER GOLDSMITH
TRISTRAM SHANDY By LAURENCE STERNE

ONE VOLUME OF "VOYAGES AND TRAVELS,"

CONTAINING

ANSON'S VOYAGE ROUND THE WORLD
BLIGH'S MUTINY OF THE BOUNTY
STEPHENS S TRAVELS IN EGYPT ARABIA PETRÆA, AND THE HOLY LAND
STEPHENS'S TRAVELS IN GREECE, TURKEY, RUSSIA, AND POLAND

These Volumes may be had separately, very neatly bound in cloth with the contents lettered on the back price 10s 6d each

Those parties who have taken in the different works as they were published, and who wish to bind them according to the above arrangement, may be supplied with the title-pages gratis, through their bookseller, and with the cloth covers at 1s each

The fore wings also exhibit near the anal angle several additional white spots, and the anal edge of the hind wings is pale bluish

A remarkable variety in which the white spots on the wings are nearly effaced, the white band being also entirely or nearly obliterated, as well as the dark mark on the under side, is figured by the Rev W T Bree, in Loudon's Magazine of Natural History, vol v p 667 The specimen was taken near Colchester, by Dr MacLean Mr Ingall also possesses a similar specimen from the same neighbourhood

The caterpillar is green, with the head, legs, and dorsal tubercles reddish It feeds on the honeysuckle A careful figure of it, from an original drawing in the collection of M Boisduval, is given in the Crochard edition of the Règne Animal, Ins pl 137, fig 4 The chrysalis has the head beaked and bifid, and a very large and prominent dorsal appendage It is brownish or green with golden spots

The butterfly appears in July, and is a rare species, although formerly more abundant, it appears widely distributed over the southern parts of the kingdom Near Peterborough, near Ipswich, Hartley Wood, Essex, near Rye, Coombe Wood, near Finchley, Birchwood, Kent, Enborne Copse, Berks, New Forest "abundantly in woods near Winchester, also a specimen in the Isle of Wight" Rev W T Bree MSS

"The graceful elegance displayed by this charming species when sailing on the wing is greater perhaps than can be found in any other we have in Britain There was an old aurelian of London so highly delighted at the inimitable flight of Camilla, that long after he was unable to pursue her he used to go to the woods and sit down on a stile for the sole purpose of feasting his eyes with her fascinating evolutions" (Haworth, Lep Brit p 30)

The remaining British species belonging to the family Nymphalidæ constitute a group of very great extent, the number of the European species being considerably greater than one-third of the whole of the Diurnal Lepidoptera of Europe They form the genus Hipparchia of Fabricius (together with part of his genus Melanitis, or the subsequently-named genus Satyrus of Latreille, or Erebia of Dalman) By Boisduval they are formed into a distinct tribe, Satyrides (Satyridæ, Swainson), and by Hubner into a stirps named Dryades, whilst by Dr Horsfield they are considered as the types of one of the five primary divisions of the Diurnal Lepidoptera, most of the other Nymphalidæ belonging to one of his other primary divisions

These butterflies are of the middle size, with the wings ornamented beneath with eye-like spots, and entire or scalloped, but never angulated, nor with the outer margin of the fore wings concave They have the discoidal cell of the hind wings closed, whilst the base of one or more of the longitudinal veins of the fore wings is dilated and vesiculose The arrangement of these veins offers no difference between this genus and the other Nymphalidæ The two fore legs are minute and rudimental in both sexes, the antennæ are terminated by a curved club, which is generally slender and spindle-shaped, but in a few species very distinct, the eyes are either naked or hairy, the palpi are not close together, the under side being clothed with long hairs But the most characteristic mark of distinction consists in the form of the caterpillars, which are attenuated at the posterior extremity, and pisciform, with the tail terminated by a small fork, the body is destitute of spines and is generally pubescent, with the head more or less rounded, and sometimes heart-shaped The chrysalis is but very slightly angulated, and almost destitute of prominent tubercles

The species feed exclusively upon the different species of grasses, and are consequently widely dispersed almost over the whole globe

The relations of these insects with the other tribes of Diurnal Lepidoptera are very interesting. In the habit of the caterpillars as well as of the imago, as suggested by Mr. Curtis, they approach Pieris (Pontia); but the supposed resemblance with the Melitææ appears to me to be very slight. Boisduval has more correctly indicated the relation of their larvæ with those of Morpho and Brassolis, as well as with Apatura, and of the imago with Biblis.

The distribution of these insects has hitherto been but little attended to. By Mr. Curtis (who has in these insects alone departed from his usual plan of giving only one illustration of each genus) they are formed into a single genus, divided into two groups, from the hairy or naked eyes. Mr. Stephens, by a more careful examination of the structure of the different species, has divided the genus into five sections, in the following manner:—

A.—Eyes pubescent; wings, especially the posterior, more or less denticulated; palpi moderately hairy; frequent woods, lanes, and highways. *Ægeria, Megæra.*

B.—Eyes naked; the wings, especially the posterior, more or less dentated; palpi moderately hairy; frequent heaths, commons, and meadows; subdivided from the form of the club of the antennæ and of the wings. *Semele, Galathea, Tithonus, Janira, Hyperanthus.*

C.—Eyes naked; anterior wings entire, rounded; posterior dentated; palpi hairy, terminal joint short, obtuse; frequent mountainous districts or swampy heaths. *Ligea, Blandina.*

D.—Eyes naked; wings elongate, pilose, entire; palpi very hairy; frequent mountainous districts. *Cassiope.*

E.—Eyes naked; wings entire; palpi slender, moderately hairy; terminal joint very long, acute; frequent boggy heaths and marshy places in mountain districts. *Polydama, Davus, Hero, Ascanius, Pamphilus.*

M. Boisduval, in his beautiful "Icones des Lépidoptères," has divided these insects into four genera:—Arge (the group typified by Galathea); Erebia, corresponding with the mountain groups, (Stephens' sections C. and D.); Chionobas, an Arctic group; and Satyrus, formed of the remainder, and divided into nine races.

M. Duponchel, in a memoir published in the Annals of the French Entomological Society for 1833, has regarded these insects as constituting but a single genus, and as divisible into nine groups, characterised by the variations in the dilatations at the base of the veins of the wings (a character entirely neglected by our English authors), and the form of the antennæ. The following are his groups, with the names of the English species belonging to each:

1. GRAMINICOLES, Galathea. 2. ERICICOLES, Phædra. 3. RUPICOLES, Briseis and Semele. 4. HERBICOLES, Janira and Tithonus. 5. VITICOLES, Megæra and Ægeria. 6. RAMICOLES, Hyperanthus. 7. DUMICOLES, Hero, Ascanius, Iphis, Davus, and Pamphilus. 8. ARCTICOLES (no British species). And 9. ALPICOLES, Cassiope, Blandina, and Ligea.

The great extent of the group, and the variation in the characters noticed above, to which others of still greater importance (but which have been neglected by preceding authors) must be added, induce me, after much consideration, to break up the old genus Hipparchia, instead of treating it as I have done the Fritillaries and Vanessæ, and to adopt a plan of distribution intermediate between those of Boisduval and Duponchel. The genera Arge (Graminicoles, Dup.), Chionobas (Arcticoles, Dup.), and Erebia of Boisduval (Alpicoles, Dup.), appear to me to be natural groups, although there is a marked difference in the form of the wings of Blandina and Cassiope, belonging to the last-mentioned group; but the genus Satyrus of Boisduval is a complete magazine, comprising species with naked and hairy eyes, smooth and pubescent larvæ, one, two, or three of the veins dilated at the base, &c. From this mass I therefore propose to detach the Viticoles of Duponchel, having, in addition to his characters, the eyes hairy, and his Dumicolæ, additionally distinguished by the glabrous larvæ and very long terminal joint of the palpi. I thus leave together all the species which have the anal vein of the fore wings not swollen, the mediastinal and median alone being more or less dilated. This group will, therefore correspond with Mr Stephens' section B, after the removal of Galathea.

These groups are further confirmed by the variations in the structure of the fore feet in the different sexes, a character which has been neglected by all previous authors, except Mr Curtis, who, without noticing the variations or even the sexual distinctions in this part, merely describes the fore tarsi of the genus as four-jointed, while Zetterstedt states that the males have the fore legs pilose, and the females almost naked, without mentioning the difference in the number of their joints or in their formation.

DESCRIPTION OF PLATE XVII

INSECTS.—Fig 1 Arge Galathea (the Marbled White Butterfly) 2 Showing the under side 3 The Caterpillar
4 The Chrysalis
„ Fig 5 A dark variety of H Galathea 6 Showing the under side
„ Fig 7 Lasiommata Ægeria (the Speckled Wood Butterfly) 8 Showing the under side 9 The Caterpillar
10 The Chrysalis

Fig 1, Arge Galathea, from an English specimen in Mr Westwood's cabinet, is exactly identical with several which I took in Italy, in the neighbourhood of Civita Vecchia, and yet the varieties found on the Continent (by many considered distinct species) are almost numberless. The present it would seem however, is the type of the species, as it is by far most abundant and constant. In England the species does not seem so prone to variation, but several varieties have nevertheless occurred, one of the most remarkable of which is the dark one, No 5 and 6 of this Plate, first figured and described by the Rev W T Bree in the Magazine of Natural History. The dark markings of this handsome species are generally described as *black*, but they are in fact a deep rich *brown*. This want of exact accuracy in entomological descriptions exists also in other species; for instance the dark portion of the apex of the anterior wings of Cynthia Cardui, generally described as black, is also a full, rich brown, and this is not the case in pale specimens only, but on the contrary is still more evident in the most strongly marked individuals. This may be at once very plainly illustrated by comparing the markings in question with those of a really pure black in V Urticæ, or P Machaon, or many others. The absence of accurate discrimination in describing dark colours is perhaps not of much consequence, and yet, as it might so easily be corrected, it appears desirable that attention should be called to it. I will mention one instance where in consequence of this defect, the description might almost apply to some other insect. In the Naturalists' Library, Entomology, vol iii, Vanessa Atalanta is described as having "the upper side of a deep black with a deep silky gloss," &c &c. The rest of the description is accurate, but that portion which makes the entire ground deep black, is quite the contrary. The fact is that the ground of the posterior wings is a fine rich silky brown, and the anterior wings are of the same colour from the base as far as the red band, beyond which the ground colour is intense violet, approaching to black but not black.

L Ægeria is from a very brightly-marked specimen in the collection of Mr Westwood, very much brighter than the individuals which I have been in the habit of taking on the Continent, the light markings of which, instead of being of a light, clear straw colour, are generally of a dusky orange. The Caterpillars and Chrysalides are from Godart. H N H

ARGE*, SCHRANK

This genus is distinguished by having the eyes naked; the antennæ elongated, with a long and slender spindle-shaped club gradually commenced; the palpi are composed of attenuated joints, the last of which is distinctly pointed and naked at the tip, the under side of the preceding joints clothed with long hairs; the hind wings are dentated; the mediastinal vein alone of the fore wings is vesiculous at the base, both above and beneath. The

* Probably derived from Ἀργῆς, *albus*, from the prevailing white colour, or else from Ἀργός *otiosus*, from the weak flight of the insect

fore legs, in both sexes, are so extremely minute as not to be visible amongst the hairs upon the breast, those of the female are still more minute than those of the male, but shorter and thicker in proportion to their size, they are alike clothed with scales, and the tarsal portion is not articulated

The larvæ have the body slightly thickened in the middle, cylindric attenuated to the tail, which is forked The chrysalis is destitute of tubercles

The perfect insects are found in grassy places in woods

This genus is exclusively composed of the species (numerous on the Continent, but of which only one has been found in England) which have the ground-colour of the wings white, marked with black spots, hence they are called by the French Leucomélamens, White Satyrs and Semi-devils They constitute the group Graminicoles of Duponchel M Lefebvre has published a valuable memoir on this group in the first volume of the Annals of the Entomological Society of France

---

SPECIES 1 —ARGE GALATHEA THE MARBLED WHITE, OR MARMORESS

Plate xvii fig 1—6

SYNONYMES — *Papilio Galathea*, Linnæus, Haworth Lewin Papil pl 28 Donovan Brit Ins vol viii pl 258 Wilkes, pl 100 Harris Aurelian, pl 11 fig g—k

Hipparchia Galathea Leach Stephens, Curtis Duncan Brit Butt pl 23, fig 1
Arge Galathea Boisduval Hubner
Satyrus Galathea Latreille, Duponchel

This singularly-marked butterfly, which from the contrasts of its colours was called the ' half-mourner" by our early aurelians, varies in the expanse of its wings from 2 to $2\frac{1}{4}$ inches Its colours, which are yellowish white and almost black are distributed in nearly equal proportions over the wings The ground colour on the upper side is almost black with one large whitish oval spot near the base of the costa, succeeded by four long whitish patches, the two middle ones being nearest the apex of the wings and smaller than the others, between these and the apex are two smaller white spots, and there is a row of white submarginal spots The hind wings have a large oval whitish spot near the base, succeeded by a very broad bar of the same colour, and with a row of submarginal white crescents varying in size

The markings on the under surface of the wings are nearly similar, except that the blackish markings are much paler, especially in the hind wings, where they are marorated with buff Moreover, the fore wings have a small black eye, with a white centre near the tip, and the posterior wings have five eyes placed just above the white submarginal crescents (the third crescent from the outer angle of the wings not having an eye), and the eye nearest the anal angle being doubled

The female differs in being of a larger size and in having the under surface of the wings of a yellower hue than in the males Some specimens in the British Museum are so strongly characterised in this respect, that I at first thought it probable they constituted a distinct species Varieties of this species are described both accidental and apparently permanent Of the former, one of the most singular is represented in our pl 17 fig 5, 6, from a specimen taken near Dover, and kindly communicated to us by the Rev W T Bree, who has published a notice of it in Loudon's Magazine of Natural History, vol v p 335 A similar variety is also figured by Ernst, Pap d'Europe, 1, pl 30, fig 60 The black marks in this variety are very greatly suffused over the largest portion of the wings An apparently permanent variety, with pale yellowish brown markings in lieu of the black ones, is described by Stephens The Arge Procida of Herbst is esteemed by Boisduval also as a local variety, owing to climate the black markings in this are much more extended, especially on the upper surface of the wings In like manner, Boisduval regards the Arge leucomelas of Esper as another local variety in which the hind wings on the under side have the black markings replaced by so very pale a shade of buff as to cause the wings to appear almost white the eyelets being always absent

The caterpillar is yellowish green, with a darker line down the back and on each side It feeds on the cat s-tail grass

The perfect insect appears in June and July It especially frequents damp open places in woods, and although local, it seems to be distributed over the greater part of England it has not, however, been found in Scotland

LASIOMMATA*, Westwood

This genus is at once distinguished from all the other Hipparchides by having the eyes thickly clothed with hairs, in addition to which the palpi are very slender, moderately clothed to the tip beneath with long hairs, the terminal joint being very short, the wings, especially the posterior pair, are denticulated, and considerably varied, the fore wings with one and the hind ones with five or six eyes, the antennæ are straight, distinctly annulated with black and white, and with the club pyriform the mediastinal and median vein are more or less swollen at the base, the anal one being simple The fore legs, although considerably smaller than the intermediate ones, are yet very conspicuous, they are of equal length in both sexes, but those of the males are comparatively slender and more densely clothed with long slender hairs, the tarsal portion in the male is simple, but in the female it is broader and articulated with several short strong spines at the tips of the joints on the under side, the larva of L Megæra is elongated, villose, with two short points at the tail, and the pupa is short, thick, with small angular points, and two points at the head, it is suspended by the tail The chrysalis of L Mæra, according to M Marloy, is suspended by the tail in the open air, it is naked and angular with two points on the head, and with broad brown bands on the wing-covers This genus corresponds with the first section of Hipparchia, of Curtis and Stephens, and with Duponchel s fifth group Vicicoles, the species being stated to occur in the neighbourhood of habitations Stephens more correctly states that they frequent woods, lanes, and highways They form Hubner's two groups, Pararge and Dira

SPECIES 1 —LASIOMMATA ÆGERIA THE SPECKLED WOOD, OR WOOD ARGUS BUTTERFLY

Plate xvii fig 7—10

Synonymes —*Papilio Ægeria*, Linnæus, Haworth, Lewin Papil pl 19 Donovan Brit Ins 11, pl 498 Wilkes Engl Butt, pl 103 Harris Aurelian, pl 41 fig f—i Sepp I, tab 6

Hipparchia Ægeria, Fabricius Ochsenheimer Leach, Stephens, Curtis, Duncan, Brit Butt pl 23 fig 4

Satyrus Ægeria, Latreille Boisduval Duponchel

Pararge Ægeria, Hubner (Verz bek Schmett)

This butterfly varies in the expanse of its wings from 1½ to 2 inches The ground colour of the wings on the upper side is brown The fore wings are marked with a number (ten or eleven in the strongest marked

* Derived from the Greek λασιος *hirtus*, and ὄμμα, *oculus* from the hairiness of the eyes

individuals) of pale buff patches of variable size, placed irregularly, the one nearest the apex of the wing being ornamented with a black eye, having a white dot in the centre, the hind wings are more sparingly marked with pale patches, but in the centre towards the margin they have three larger eyes placed in a bar of pale buff On the under side, the brown colour in the fore wings is more clouded, the apex being much paler with the eye near the tip whilst the hind wings are more varied with lighter and darker undulations, the outer angle being paler, and a row of six white dots, varying in size near the hinder margin (which has sometimes a purplish tinge), the larger ones replacing the others of the upper side There is great variation in the size and number of the pale spots, as well as of the clouding of the under surface of the wings, and the females are generally ornamented with larger and more numerous spots than the males

The caterpillar of this species is green, with white longitudinal lines and a spined tail It feeds upon grasses preferring the common couch grass, and is found in March, May, and June, the perfect insect appearing in April, June, and August, there being several broods in the course of the year It delights in lanes and glades of woods, and is a common species, occurring from Dover to the north of Scotland

DESCRIPTION OF PLATE XVIII

INSECTS —Fig 1 Lasiommata Megæra, male 2 The female 3 Showing the under side 4 The Caterpillar 5 The Chrysalis
" Fig 6 Hipparchia Semele, male 7 The female 8 Showing the under side 9 The Caterpillar 10 The Chrysalis
PLANTS —Fig 11 Bromus sterilis (barren Brome-grass)

The insects figured on this plate are all from English specimens in Mr Westwood's, or my own collection and the caterpillars are from Godart

Hipparchia Briseis a common species on the Continent, having been on one occasion discovered in England, I shall take the opportunity of giving a figure of it in this work, and intended that it should have appeared in juxtaposition with H Semele on the present plate But the space does not admit of it, and it will therefore be given in the next H N H

SPECIES 2.—LASIOMMATA MEGÆRA THE WALL BUTTERFLY

Plate xviii fig 1—5

SYNONYMES — *Papilio Megera* Linnæus, Lewin Papil pl 21 Donovan Brit Ins 8, pl 279 Supp v 2, pl 2, 3 Wilkes, 53, pl 102
Papilio Mægera, Haworth
Hipparchia Megæra Ochsenheimer, Leach Stephens, Curtis
Satyrus Megæra Latreille, Boisduval, Duponchel
Dira Megæra, Hubner (Verz bek Schmett)
Papilio Mæra, Berkenhout Harris Aurelian pl 27 fig a—g

This pretty butterfly varies in the expanse of its wings from 1½ to nearly 2 inches The ground colour of the upper surface of the wings is of a fulvous yellow, with several transverse irregularly undulating brown bars, the base of the hind wings being also brown, as well as the margin of all the wings Near the tip of the fore wings is a large black eye with a white pupil, and the hind wings have a row of from three to five black eyes varying in size, the middle ones also having a white pupil The male differs in having a broad oblique brown bar extending across the middle of the hind part of the fore wings On the under side, the fore wings are nearly marked as above, except that the brown bars are more slender, and the broad oblique bar of the male is wanting The ocellus is surrounded by a brown ring, and accompanied by another minute ocellus The under wings are beautifully freckled with ashy and brown, with many waved darker marks, forming a broadish curved bar across the middle of the wings, beyond which is a row of six beautiful eyes, that at the anal angle being double, succeeded by a row of darker waves

The caterpillar is slender, pubescent, and of a light green colour, with darker lines on the back and sides It is found at the beginning of May and August, and feeds on grasses The imago appears in July and August and is a common and widely-dispersed species, frequenting lanes and road-sides, delighting to settle on walls (whence its ordinary English name), flying off when approached and settling at a short distance again to be disturbed at the approach of the passer-by

PAPILIO MÆRA (Linnæus), a species closely allied to the preceding, and placed in the same subgenus Dira by Hubner, has been introduced into the English list of butterflies, in consequence of Linnæus having erroneously referred Wilkes' figure of Megæra to Mæra, which evidently induced Berkenhout to give Megæra under the name of Mæra Although similar to Megæra in its markings, it is at once distinguished by the more distinct club of the antennæ, a character pointed out by Zetterstedt

---

HIPPARCHIA*, FABRICIUS

This genus is distinguished from the preceding with which it agrees, in having the mediastinal and median veins above more or less dilated at the base, in the naked eyes The wings are generally considerably variegated, and more or less denticulated, especially the hinder pair and the palpi moderately hairy The antennæ vary in the construction of the club, which in some species is long, slender, and fusiform, and in others abrupt and broad The fore legs are of comparatively moderate length, and distinctly visible in both sexes, those of the males being much more densely clothed with hair, and those of the females rather larger The tarsal portion is simple in the males, but articulated in the females, without however the short spines at the tips of the joints beneath observed in the Lasiommatæ† This genus comprises the greater part of Stephens section B of Hipparchia, and with Duponchel's Ericicoles (Phædra, &c), Rupicoles (Briseis, Semele, &c) Herbicoles (Janira and Tithonus, &c), and Ramicoles (Hyperanthus, &c)

The larvæ are conical, with the head round, and the tail bifurcate, they are marked with several longitudinal black stripes M Maloy, who has published a short notice on the larvæ of these insects in the Annals of the Entomological Society of France for 1838, mentions that the chief cause why these larvæ are so seldom met with is, that they conceal themselves and remain inactive during the day, but come forth to feed by night, when they may be found in great numbers with the help of a lamp The caterpillars of Briseis and Semele form large cocoons under ground, composed of grains of earth fastened together with a little silk, their chrysalides are short ovoid glabrous, contracted, with the head obtuse and the tail pointed Janira differs from the preceding, in having the chrysalides naked, angular with the head bifid, and suspended head downwards The larva of Tithonus, according to Boisduval, has the hairs of the body bifid

---

* Unmeaningly derived, by Fabricius, from the Greek Ἱππαρχία *præfectura equitum*

† The tarsi above described are those of Semele Those of Janira are shorter (although conspicuous) and very slightly pilose, with the tarsal portion in the males short and slightly compressed, but longer in the females, and articulated In other respects they are nearly alike in size and appearance In Tithonus they are very minute in both sexes, but rather larger in the females, and very slightly hairy in the tarsal part, more elongated than in the male, and thick at the tip The same minuteness of size occurs also in Hyperanthus, thus confirming the subdivisions established on other characters

SPECIES 1.—HIPPARCHIA SEMELE. THE GRAYLING.

Plate xviii. fig. 6—10.

SYNONYMES.—*Papilio Semele*, Linnæus. Lewin Papil. pl. 17. Donovan Brit. Ins. pl. 250. Haworth. Harris Aurelian pl. 44, fig. d. e. *Hipparchia Semele*, Ochsenheimer, Leach, Stephens. Duncan Brit. Butt. pl. 22, fig. 1, 2.

Satyrus Semele, Latreille, Boisduval. Duponchel. *Eumenis Semele*. Hubner.

This is the largest of our common British Hipparchiæ, measuring from 2¼ to more than 2½ inches in expanse. The fore wings on the upper side are of a dull brown colour, with a broad interrupted bar of various size near the extremity, in which are two black eyes; the hind wings are brown at the base with a brighter coloured bar near the margin, having a single black eye with a white centre near the anal angle; on the under side, the fore wings are darker at the base, with the extremity yellowish or pale buff, terminated by a narrow dusky margin. The two ocelli are here distinct, the anterior one being largest; the under wings on this side are marked with very numerous short, slender, transverse, white, brown, and black streaks; the basal half is darkest, and is terminated by a very irregular broad paler bar; near the anal angle is a nearly obsolete eyelet.

The markings vary greatly in size as well as in the intensity of their colours; and the females have the marks and eyes larger, but paler.

The caterpillar is green or grey, with the belly and legs brownish; it is rather more than an inch long; its body is thick, hard, and conical, with five blackish longitudinal lines, the dorsal one being the darkest. It forms a cocoon in the earth, according to M. Mulloy. The butterfly, which appears in July, is rarer than the preceding, owing to its preferring certain localities, such as heaths (Newmarket, Gamlingay, and Salisbury Plain, for example), and rocky places, such as Arthur's Seat near Edinburgh; and stony places, near Durham and Castle Eden Dene. Mr. Wailes also observed it frequently on the sea-coast, near South Shields, where the magnesian limestone occurs, although not found on the opposite side of the Tyne, where there is no limestone.

PAPILIO PHÆDRA of Linnæus appears to have been introduced by Turton as a British species without sufficient authority. It has all the wings on the upper side of a deep uniform brown, the fore ones with two large ocelli, and the hinder ones with a single minute one near the anal angle. It measures 2½ inches in the expanse of its wings.

PAPILIO ALCYONE of Esper has also been erroneously given as a native of Scotland, by Stewart, who mistook Blandina for it. It has the wings brown with a whitish bar, the anterior having two ocelli on each side, and the posterior beneath marbled, with a white angular bar and a single ocellus.

DESCRIPTION OF PLATE XIX.

INSECTS.—Fig. 1. Hipparchia Briseis, female. 2. Showing the under side.

" Fig. 3. Hipparchia Tithonus (the large Heath Butterfly), male. 4. The female. 5. Showing the under side. 6. The Caterpillar. 7. The Chrysalis.

PLANTS.—Fig. 8. Poa pratensis (Common Meadow Grass).

The Hipparchia Briseis figured on this plate seems to have a fair claim to be considered British, as it was raised from the caterpillar found in the neighbourhood of London; of which I much regret not being able to give a drawing. It will be seen, on comparison with the figure in the previous plate, that the insect is closely allied to the common H. Semele. H. N. H.

With numerous Woodcuts, price 6s

THE LADIES' COMPANION TO THE FLOWER-GARDEN.

Being an Alphabetical Arrangement of the Ornamental Plants usually grown in Gardens and Shrubberies, with full directions for their Culture

BY MRS LOUDON.

Complete in One Volume 4to, price 2l 2s cloth, or 2l 10s half bound morocco, gilt edges

THE LADIES' FLOWER-GARDEN OF ORNAMENTAL ANNUALS.

BY MRS LOUDON

ILLUSTRATED WITH FORTY-EIGHT CAREFULLY COLOURED PLATES

Containing upwards of 300 Figures of the most showy and interesting Annual Flowers

---

THE IMPERIAL CLASSICS.

Just published, Part VII, price Two Shillings, of

BISHOP BURNET'S HISTORY OF THE REFORMATION,

WITH HISTORICAL AND BIOGRAPHICAL NOTES

To be completed in Thirteen or Fourteen Parts

Just published, No XI, price 2s 6d

THE LADIES' FLOWER-GARDEN OF ORNAMENTAL BULBOUS PLANTS.

BY MRS LOUDON.

Each Number contains Three Plates, demy 4to size, comprising from Twelve to Twenty Figures accurately coloured from nature
The whole will occupy about Twenty Numbers.

A PRESENT FOR THE YOUNG.

Handsomely bound in cloth, gilt, price 4s 6d

TRUE TALES, FROM FROISSART.

ILLUSTRATED WITH SIXTEEN WOODCUTS

THE LADIES' MAGAZINE OF GARDENING.

BY MRS LOUDON.

In Monthly Numbers, Royal 8vo, with Coloured Plates. The First Number was published on January 1st, 1841, price 1s 6d

---

Just published, No. II, price 1s

THE SUBURBAN HORTICULTURIST.

BY J C LOUDON, F L S, H S., &c

To be completed in Twelve Octavo Numbers

(To be completed in about Sixteen Numbers.)

No. VIII.] [PRICE 2s. 6d.

BRITISH INSECTS AND THEIR TRANSFORMATIONS.

BRITISH BUTTERFLIES

AND

Their Transformations.

ARRANGED AND ILLUSTRATED IN A SERIES OF PLATES

BY H. N. HUMPHREYS, ESQ.

WITH CHARACTERS AND DESCRIPTIONS

BY J. O. WESTWOOD, ESQ., F.L.S.,

SEC. OF THE ENTOMOLOGICAL SOCIETY, ETC. ETC.

LONDON:

WILLIAM SMITH, 113, FLEET STREET.

MDCCCXLI.

SPECIES 2—HIPPARCHIA BRISEIS

Plate xix fig 1, 2

SYNONYMES—*Papilio Briseis*, Linnæus, Fabricius Erns. Pap 1, pl 21, fig 36, a—d Naturforscher 10, tab 2, fig 3, 4
Hipparchia Briseis, Ochsenheimer

This fine species measures about 2½ inches in the expansion of the wings, which are denticulated, and of a brown colour, having a greenish gloss. The anterior have an interrupted row of pale buff spots near the outer margin, which are extended across the hind wings moreover the fore wings are marked with two black ocelli with a white centre The under surface of the fore wings is more varied with brown and buff than the upper, especially towards the base, and is marked on the costa with two blackish spots, and the tip is freckled The hind wings also on this side are freckled with short transverse streaks A very irregular paler bar runs across the middle of the wing, succeeded by a darker waved one in which are several rudimental ocelli

The caterpillar is smooth, thick, and greyish-coloured, conical, with five longitudinal dark lines, the dorsal one being the darkest the head is round and red It forms a cocoon under ground, according to M Marloy

We have introduced this common Continental species for the first time as an English insect, a specimen having been reared by A Lane, Esq, from the larva which was found feeding on grass near Newington The perfect insect was exhibited at the meeting of the Entomological Society on the 7th of October, 1839, the larva having been captured on the 11th of August preceding.

SPECIES 3—HIPPARCHIA TITHONUS THE GATE-KEEPER, OR LARGE HEATH BUTTERFLY

Plate xix fig 3—7

SYNONYMES—*Papilio Tithonus*, Linn (Mantissa), Lewin Pap pl 22 Harris Aurelian, pl 44, fig f, g
Hipparchia Tithonus, Ochsenh, Steph, Curtis, Duncan, Brit Butt. pl 23, fig 2, 3
Pyronia Tithonus, Hubner (Verz bek Schmett)
Papilio Tithonus, Villars (non his Pilosellæ)
Papilio Herse, Hubner, Pap
Papilio Phædra, Esper
Papilio Pilosellæ, Fabricius, Haworth, Donovan, Brit Ins v 12, pl 405

This common butterfly varies from 1¾ to nearly two inches in the expanse of its wings, the ground colour of which on the upper side is of an ochre yellow, with a broadish brown margin The base of all the wings is also brown, near the apex of the fore wings is a large black eye, in which are two small white dots, near the anal angle of the hind wings is also a nearly obsolete eye, more strongly marked in the female The male is distinguished by its smaller size, more obscure colouring, and by having a broad brown oblique patch in the middle of the posterior disc of the fore wings The fore wings on the under side are coloured as on the upper, except that the brown patch is wanting in the males the hind wings, on the contrary, are of a golden brown at the base and margin, with an irregular waved greyish-buff band running across the middle, having a brown patch near the outer angle in which are two small eyes, and another patch and ocellus towards the anal angle, sometimes accompanied by one or two small white ocelli The size of these ocelli, as well as their number, varies in different specimens

The caterpillar is greenish, pubescent, with a reddish line on each side, and a brownish head It feeds on the annual meadow-grass, and also (according to Haworth) on the Hieracium Pilosella It is found in this state in the beginning of June, and the butterfly appears in the middle of July It is a very abundant species, frequenting pasture lands and lanes throughout the kingdom

DESCRIPTION OF PLATE XX

INSECTS.—Fig. 1. Hipparchia Janira, male (the Meadow-Brown Butterfly). 2. The female. 3. Showing the under side. 4. The Caterpillar. 5. The Chrysalis.

" Fig. 6. Hipparchia Hyperanthus (the Ringlet Butterfly). 7. Showing the under side. 8. The Caterpillar. 9. The Chrysalis.

PLANT.—Fig. 10. Poa annua (Annual Meadow Grass).

H. Hyperanthus has been grouped with H. Janira upon this plate to display the singular resemblance of the male insect of the two species on the upper side, while they are so strikingly different beneath. H. N. H.

SPECIES 4.—HIPPARCHIA JANIRA. THE MEADOW BROWN BUTTERFLY.

Plate xx. fig. 1—5.

SYNONYMES.—*Papilio Janira*, Linnæus, (male,) Turton, Stewart. *Hipparchia Janira*, Ochsenheimer, Stephens, Leach, Curtis, Duncan, Brit. Butt. pl. 24. fig. 1, 2.

Papilio Jurtina Linnæus, (female,) Lewin Pap. pl. 18. Donovan, I. pl. 320. Haworth, Harris Aurelian, pl. 32, fig. a—e. *Papilio Hyperanthus*, Wilkes 53, pl. 101. Albin, pl. 53 fig. a—c. *Epinephile Hyperanthus*, Hubner (Verz. bek. Schmett.)

This most abundant species varies in the expanse of its wings from $1\frac{3}{4}$ to 2 inches. As its English name imports, the prevailing colour of the wings on the upper side is obscure brown or almost black, especially in the males. Both sexes have a small black eye, with a white centre, placed on a small fulvous patch near the tip of the fore wings, and the female has a large fulvous patch beneath the ocellus, which is sometimes also slightly visible in the males. On the under side, the wings are brighter coloured, the fore ones being dark orange yellow, lighter beyond the middle, and with the margin pale brown. The ocellus near the apex is also here present, the basal half and the margin of the hind wings are tawny brown, separated by a broad irregular paler bar in which are from one to three minute dark dots. The markings of this species, however, greatly vary in size, as well as occasionally in colour, and the ocellus of the fore wings is sometimes without and sometimes with two white dots, occasionally also it is accompanied by one or two black spots beneath, as in fig. 2.

The caterpillar is pubescent, green, with white longitudinal lines, and the tail is forked. It feeds on several species of grass, especially Poa pratensis. The chrysalis is naked and angular, suspended by the tail, with two sharp points at the head.

The butterfly, which is to be found in every meadow and grassy lane, is one of the commonest of our English species, and occurs all over the kingdom. Mr. Knapp, the author of the pleasing Journal of a Naturalist, notices that it appears but little affected by the diversity of seasons, being equally copious in damp and cheerless summers as in the driest and most arid ones. Indeed, in 1826, which was exceedingly parched, the number of these butterflies was so great as to attract the attention of different persons.

Linnæus mistook the sexes of this butterfly for different species, but their specific identity has long been unquestionably established. In such cases the name given to the male specimens is retained instead of that of the female.

SPECIES 5.—HIPPARCHIA HYPERANTHUS. THE RINGLET BUTTERFLY.

Plate xx. fig. 6—9.

SYNONYMES.—*Papilio Hyperanthus*, Linnæus, Lewin Pap. pl. 20. Donovan, Brit. Ins. 8. pl. 271. Haworth, Harris Aurelian, pl. 25 fig. d—h (not of Wilkes). *Hipparchia Hyperanthus*, Ochsenheimer, Leach, Stephens, Curtis, Duncan.

Papilio Polymeda, Scop., Hubner, Pap. *Satyrus Hyperanthus*, Boisduval. *Enodia Hyperanthus*, Hubner (Verz. bek. Schmett.)

This plain-coloured butterfly varies in the expanse of its wings from $1\frac{1}{2}$ to nearly 2 inches. The upper surface of all the wings is dark brown, without any shade or mark except one or two small and more or less distinct ocelli near the hind margin of the fore wings, and with several similar ocelli near the margin of the hind wings On the under side, the ground colour of the wings is rather paler, but uniform, whilst the ocelli of the upper side are here repeated in the hind wings, but greatly increased in size, there being generally two near the outer angle, and three between the middle and the anal angle But these ocelli vary in an endless manner, in some the ocelli being not only very large and connected together but also accompanied by smaller ocelli attached to them, whilst in others the ocelli are obliterated, the fore wings being without any spots, and the hind wings with only three minute white spots Every intermediate variety is found

The caterpillar is pubescent, of a greyish white colour, with a slender black dorsal line, and sometimes it is entirely blackish It feeds on the Poa annua, and other grasses, keeping during the day at the roots The perfect insect appears at the end of June and is found in damp grassy places about woods and lanes, and seems to be generally dispersed throughout the country

---

CŒNONYMPHA, Hubner

This genus comprises the smallest species of the Hipparchides, distinguished from all the rest by several evident characters the wings are entire and not denticulated, the anterior pair having the three veins (mediastinal, median, and anal) strongly and equally swollen, the eyes naked the palpi slender, moderately hairy with the last joint very long and acute, the antennæ annulated with grey and brown with a decided club

The fore legs (in C Pamphilus) are of comparatively moderate length, those of the males very densely clothed with long hairs, those of the females almost naked, with the tarsi long and articulated, almost resembling a perfect foot, whilst the tarsi of the males are very short and simple

The larvæ are completely glabrous and shining, thus differing from the larvæ of all the preceding species of this subfamily

This genus, to which I have applied Hubner's name (Cœnonympha), corresponds with Stephens' section E, and with Duponchel's group Dumicoles

DESCRIPTION OF PLATE XXI

INSECT —Fig 6 Cœnonympha Davus, the small Ringlet Butterfly 7 Showing the under side

Fig 3 Cœnonympha Polydama (the specimen formerly figured by Mr Stephens), showing the under side, with the cream-coloured band or fascia continuous, and not interrupted as in Fig 2

Fig 1 Cœnonympha Typhon, the Marsh Ringlet Butterfly (the specimen formerly figured by Mr Stephens as Iphis) 2 Showing the under side 4 The caterpillar of C Iphis of Godart 5 The chrysalis

Fig 8 Another specimen of C Typhon, approaching in colour on the upper side to C Pamphilus 9 Showing the under side of the same specimen, which is also paler and more indistinct than ordinary

PLANT —Fig 10 Aira Flexuosa, a common Heath-grass

The species here figured with some of their most prominent variations have been the subject of some confusion, and have been at different times subdivided into more distinct species than there are varieties in the present plate It is now the opinion of several entomologists, more particularly of Mr Stephens, who is one of the best authorities upon the subject, that they can only form two distinct species, viz C Davus and C Typhon or Polydama—the continuity or interruption of the cream-coloured band or fascia on the under side of the hind wings, with other trifling differences of marking, being no longer considered specific distinctions These two species may be generally distinguished by the following characteristics —C Davus always has the little rings more or less defined on the upper side, and is of a dull brown with a slight inclination to grey, on the under side, the markings are very strong and perfect, and in the darker parts inclining to an olive green C Typhon or Polydama has the little rings very slight, and in some cases altogether deficient on the upper side, whilst the ground colour is somewhat paler, and instead of inclining to grey, inclines rather to tawny, and on the under side all the markings are paler and less distinct Davus is found in Lancashire, more particularly in the neighbourhood of Manchester, where Polydama is seldom or never seen whilst Polydama is confined to Yorkshire, Cumberland, and South Wales Notwithstanding these distinctions both of markings and locality, I should myself feel almost tempted still further to simplify the matter, and to refer the two reputed species to one, the different localities being rather an argument in favour of uniting the species than separating them, for it is well known that the insect of the mountain and the same insect of the plain beneath have always different characters sufficiently constant for specific distinction if there were not abundant general evidence of identity

The specimens figured in the present Plate, seem sufficient evidence that they may be all referred to one species, for though there is a wide difference between Typhon fig 9, and Davus fig 7, yet the connecting links are not wanting and indeed all the intermediate gradations might be selected from any hundred specimens indiscriminately taken Davus seems the type of the genus there, all the markings are complete, distinct, and unclouded in Polydama fig 3, they are already somewhat paler and less defined in Typhon, fig 2, the broad band or fascia on the hind wing is broken up into two irregular marks, while in fig 9 the whole is fainter and some of the marks have disappeared altogether

As the caterpillars of these Butterflies are at present unrecorded in this country, I hope some of our subscribers in the neighbourhood of Manchester, and others in Yorkshire, Cumberland, or Wales will endeavour to forward specimens to us in the course of the next season* which will decide the question, and they shall be carefully figured and described in our supplemental plates They must be sought during the fine nights of June as they only feed after sunset, and when discovered, some care must be used, as they drop from the blade of grass (on which they are feeding) directly it receives the slightest touch H N H

SPECIES 1—CŒNONYMPHA DAVUS THE SMALL RINGLET BUTTERFLY

Plate xx fig 6, 7

SYNONYMES —*Papilio Davus*, Fabricius, Haworth Jermyn (but not of Godart)
Hipparchia Davus, Ochsenheimer, Curtis, Stephens
Papilio Tullia, Hübner, Pap 213, 214
Papilio Philoxenus, Esper
Papilio Musarion, Borkhausen
Maniola Typhon, Schrank (not P Typhon, Esper, nor P Typhon of Haworth)
Papilio Hero Donovan Brit Ins 6, pl 185 Lewin pl 25 fig 5, 6 (but not P Hero of Linnæus)

This plain-coloured butterfly varies in the expanse of its wings from 1½ to 1¾ inches On the upper side they are of a brownish ochre colour, the base being rather darker and the fringe of a pale grey, parallel to the outer margin of the fore wings are the rudiments of two eyes, and occasionally with one or two smaller ones Traces also of several eyes appear near the margin of the hind wings Beneath, the fore wings have the basal half of a somewhat brighter ochre this is succeeded by a narrow irregular pale bar (narrowed to the hinder margin), the space between which and the outer margin of the wings is greyish brown, with only two larger ocelli, and sometimes with one, two, or three additional smaller ones The hind wings beneath are dark brownish grey at the base, the outer margin of which is angulated this is followed by a broad irregular whitish central bar this is succeeded by brownish grey usually ornamented with six large ocelli, having a black iris surrounded by a whitish or fulvous ring, with a small silvery central dot The markings on the under side of the wings vary in size, and the eyes in number, sometimes the whitish middle bar is interrupted in the centre and the anal ocellus is generally doubled The outer edge of the hind wings is whitish and the fringe brownish

This species is so closely allied to the P Typhon of Haworth, that it is possible that it may ultimately prove to be a local variety of it

The specimen described by Mr Haworth was captured near Manchester in the month of July Since that time it has been plentifully taken in the marshes between Stockport and Ashton near that town Trafford and White Moss, also near Manchester, and Shorn Moor, Yorkshire, have likewise been given as its localities

---

* These or any other communications may be forwarded to our publishers

PAPILIO POLYDAMA of Haworth (Polymeda of Jermyn but not of Scopoli, the last-named author having given that name to Hyperanthus) is considered by Curtis as a variety of Davus, whilst Mr Stephens at first gave it as a distinct species, but subsequently, in the Appendix to the first volume of his Illustrations, asserted it to be a variety of Typhon of Haworth It measures about an inch and a half in the expanse of its wings the fore pair of which, on the upper side, are of a greyish ochre, with two obscure blind eyes, the hind wings above are brown, but with the inner edge broadly whitish or buff, with a small obscure eyelet near the anal angle The fore wings beneath are of a brownish ochre, blackish at the base and ashy at the tips, with an abbreviated whitish fascia across the middle, between which and the outer margin are two eyes of small size the hind wings beneath have a broad basal bar of greyish brown externally dentated, terminated by a whitish irregular bar, sometimes almost interrupted in the middle, beyond which the wings are ashy, with six small eyes surrounded with a whitish ring, three of which are usually much smaller than the others The ocelli, as well as the ground-colour of the wings, vary considerably

This variety—or species as it may be considered—was first found in the county of York in the month of June, and subsequently on the 21st of July, 1809, by the Rev W T Bree, on the moors between Bala and Festiniog, North Wales, in company, however, with a specimen of Typhon, *Haw* Mr Weaver also found Polydama plentifully in North Wales in 1827, whilst he found Typhon still more profusely in Cumberland one month earlier According to Mr Wales, however, both Typhon and Polydama occur plentifully on damp heaths in Northumberland in the beginning of July Mr Curtis, on the other hand, states that it is taken near Manchester in company with Davus—"Non nostrum tantas componere lites"

Our figure 3, in plate 21, is taken from the specimen which was represented by Mr Stephens, Illustr Haust 1, pl 7 fig 3

SPECIES 2—CŒNONYMPHA TYPHON, *Haworth* THE SCARCE HEATH BUTTERFLY

Plate xxi figs 1, 2, 4, 5, 8, and 9

SYNONYMES —*Papilio Typhon*, Haworth, Jermyn Curtis

Papilio Tiphon, Esper (according to Stephens but disputed by Curtis)

Papilio Iphis (*Wien V* ?) Borkhausen, Jermyn Stephens, Illust Haust I, pl 7, fig 1, 2 (but not of Ochsenheimer, Hubner, nor Godart, according to Curtis, the species described by those authors being the Hero of Fabricius)

Satyrus Davus, Godart

Hipparchia Polydama, var, Stephens, Syst Cat and Illust Haust I App p 148

This species (if indeed it be specifically distinct from Davus and Polydama) varies from $1\frac{1}{4}$ to $1\frac{1}{2}$ inches in expanse On the upper side the wings are usually of a rusty grey or ochre colour, having the base brownish The hind wings are generally darker, and without any trace of rudimental eyes, but sometimes with distinct ocelli varying in number near the tip, on the under side the fore wings are dusky at the base, with the disc rusty ochre, followed by an abbreviated, irregular white stripe, the outer part of the wing being greenish ash, and bearing generally two (but sometimes as many as five) small eyes, which are occasionally obsolete the hind wings beneath are of a greenish-brown at the base, with an irregular, *interrupted* whitish bar beyond the middle of the wing succeeded by an ochre shade in the female, but greenish-brown in the male, and generally ornamented with six small eyes, but their number is liable to great variation The females are further distinguished by having the wings paler and more ochreous, and marked on the upper side with a large pale blotch on the disc of all the wings

The varieties of this species are very numerous, not only in the markings, but also the ground-colour of the wings. One of the most striking is represented in our Plate 21, fig 8, 9

This butterfly was first taken in June, in Yorkshire, near Beverley Many years afterwards it was again found by Mr Haworth in a marsh near Cottingham, in the same county It also occurs in Scotland, Wales, Cumberland, Northumberland, the Shetland Islands, and other parts of the North of Great Britain It was found by Mr Weaver in Cumberland, unaccompanied by Polydama, which he had found in North Wales nearly a month earlier although, according to Mr Wailes, both occur in company in Northumberland

It will be seen by the synonymes how great a confusion has prevailed as to the specific name of the insect Mr Stephens first described it under the name of Iphis, which he afterwards altered to Polydama, regarding it as identical with the Polydama of Haworth, which he had also at first considered as distinct If Polydama however, be a variety of Davus, some other name must be given to this species, and although the synonymes of Esper (to Tiphon) are disputed, and the change of the name to Typhon not perhaps strictly correct, yet I have thought it best to return to the name imposed on the species by my lamented friend and tutor in Entomology, Mr Haworth—instead of giving it a new specific name, which would otherwise have been rendered necessary

The caterpillar and chrysalis are copied from Godart's figures of C Iphis, but the synonymes of the species so named have been so confused, that we are not quite certain whether it be identical with Typhon or not

DESCRIPTION OF PLATE XXII

INSECTS —Fig 1 Cœnonympha Pamphilus (the small Heath Butterfly) 2 Showing the under side

" Fig 3 Cœnonympha Hero 4 Showing the under side

" Fig [illegible] Cœnonympha Arcanius 6 Showing the under side 7 The Caterpillar 8 The Chrysalis

" Fig 9 Orebia Cassiope (the Mountain Ringlet B) 10 Showing the under side

PLANTS —Figs 11, 12, 13 Cynosurus cristatus (Dog's tail grass) Fig 14 Melica nutans (mountain Melic-grass)

C Pamphilus is from a specimen in the possession of Mr Westwood C Hero and O Cassiope from the collection of Mr Stephens The caterpillar of Arcanius is from Godart, who describes it as feeding upon Melica ciliata which, not being an English species I have represented it upon M nutans H N H

SPECIES 3 —CŒNONYMPHA ARCANIUS

Plate xxii fig [illegible]—8

SYNONYMES —*Papilio Arcanius*, Linn [illegible] Hubner (*P Arcania*)
Hipparchia Arcanius Jermyn, Stephens, Curtis Brit Ent pl 205

This butterfly measures an inch and a half in the expanse of its wings, the fore pair of which are tawny, with the front edge and outer margin brown, having a small obscure eye near the tip, the hind wings are brown palest across the middle, with a narrow orange stripe at the anal angle, and three or four very indistinct ocelli The fore wings beneath have the front and outer edges slightly brown, with a short pale ochreous stripe towards the apex, where there is an ocellus with a black iris (sometimes accompanied by a smaller eye) The hind wings beneath are orange brown at the base, terminated in an irregular margin beyond the middle of the wing, having an ocellus with a black iris and silver pupil at the superior extremity, a broad whitish irregular band beyond is succeeded by a bright tawny margin, in which are two large and two or three small ocelli with a silver line running near the margin of the wing

Such is the description of the only specimen of this butterfly at present in any British collection—and even

its claims as a real native are questioned Mr Curtis, to whom it belongs, states that it "was captured by Mr Plastead, it is understood, on the borders of Ashdown Forest," but Mr Stephens (Illust Brit Ent Haust 4 379) doubts its authenticity It is an abundant species on the Continent appearing in the months of June and July

The caterpillar is green, with a red mouth, a darker green line down the back, a pale-yellow line down each side, and another above the feet It feeds upon Melica ciliata

A hybrid between C Arcanius and C Hero has been described and figured by Schummell (Beiträge z Entomol Schlesische Faun pl 9, f 5), having the upper side of Arcanius, and the under-side of Hero It was found in 1809 near Gerlachsdorf

SPECIES 4 —ŒNONYMPHA HERO THE SILVER-BORDERED RINGLET BUTTERFLY

Plate xxii fig 3—4

SYNONYMS —*Papilio Hero*, Linnæus, Haworth Ent Trans (not of Fabricius, nor of Donovan who figures Davus under that name) *Hipparchia Hero*, Ochsenheimer, Stephens, Curtis, Brit Ent pl 205 *Papilio Sabæus*, Fabricius *P Melibæus* Hubst

This species measures about an inch and a half in the expanse of its wings, which are of a fulvous brown colour, the fore wings being paler along the fore edge, with an orange stripe close to the posterior margin, near to which are two small indistinct orange ocelli, with brown pupils, the hind wings have also a narrow orange stripe near the outer margin, above which are four large black ocelli having minute whitish pupils, and surrounded by a broad orange ring The fore wings beneath are coloured as on the upper side, except that there is a narrow silver stripe adjoining the orange sub-marginal one, and which is also continued through the hind wings, which have an irregular whitish bar rather beyond the middle, succeeded by orange, in which are seven ocelli of various size the two nearest the anal angle being confluent and smallest, the iris being black, with a white pupil, and surrounded by an orange circle

This is one of our rarest insects, a female taken by Mr Plastead near Withyam, on the borders of Ashdown Forest, Sussex, and now in Mr. Curtis's collection being the only specimen in British collections until recently, when it appears to have been again taken in Sussex, Mr Stephens stating, in the Appendix to the Lepidopterous volumes of his Illustrations, that he had obtained a specimen from the neighbourhood of Lamberhurst

SPECIES 5 —ŒNONYMPHA PAMPHILUS THE SMALL HEATH BUTTERFLY

Plate xxii fig 1—2

SYNONYMS —*Papilio Pamphilus* Linnæus Lewin Papil pl 23, fig 3, 4 Haworth Stewart, Harris, Aurelian, pl 21, fig e—h De Geer Mém 2, pl 2, fig 3 *Hipparchia Pamphilus*, Ochsenheimer, Leach Stephens Curtis Duncan, Brit Butt pl 26, fig 3 *Papilio Nephele* Hubner, Pap

This, which is one of the commonest of our British butterflies, varies in the expanse of its wings from 1¼ to 1½ inches The wings on the upper side are of a pale tawny or fulvous colour, with the entire margins brownish the anterior pair having an indistinct ocellus near the tip, sometimes accompanied by a still smaller one, or by one or more black spots The hind wings have sometimes also an obsolete ocellus near the anal angle On the under side the fore wings are fulvous, with the base and apex ashy, a rather large ocellus being placed near the tip, having a black iris and white pupil, and surrounded by whitish The hind wings are brown at the base and ashy at the tips, with an abbreviated whitish band across the middle, beyond which are several minute, indistinct ocelli Varieties occur in which the ocelli are more or less obliterated, and in the males the dusky edging of the wing is more decided than in the females

The caterpillar feeds upon *Cynosurus cristatus*, and is found at the beginning of May and August It is greenish, with a dusky line down the back and a pale line down each side The perfect insect is found abundantly on heaths and dry pasture lands, appearing at the beginning of June and September Moses Harris also states, that there is a brood in April, making three in the course of one year

---

OREINA *, WESTWOOD

This genus is distinguished from the other British species of Hipparchides by having none of the veins of the wings dilated at the base The antennæ are slender, with a more or less globular or pyriform club The eyes are naked the palpi hairy the wings varying in shape, the anterior being either rounded or elongate, and the posterior denticulated or entire The fore feet in the males of Blandina are very small, so as not to be visible among the hairs of the breast, and very densely hairy, those of the female, on the other hand, are comparatively long, quite visible, slender, naked, and with the tarsal portion articulated

This genus, to which Boisduval inappropriately applied Dalman's generic name of Erebia (which is a synonyme of Hipparchia or Satyrus) is composed of species for the most part natives of mountainous districts hence I have applied to it a name derived from the Greek in allusion to this habitat The Continental species are very numerous and very difficult to determine Boisduval states, that they exclusively inhabit the Alpine mountains, and the mountain districts of central Europe, being but very rarely found on the plains, except where the vegetation has an alpine character They are not found on the mountains of the north of Europe (where they are replaced by the species of Chionobas), nor on the mountains of the south of Europe They constitute Duponchel's ninth and last group, named from the same circumstance Alpicoles which that author suggests may be formed into two divisions from the entire and denticulated wings indeed by Stephens they are, from this circumstance separated into two groups, forming his sections C and D of Hipparchia The species with denticulated hind wings are termed Epigea by Hubner, whilst those with entire wings are his Melampias

DESCRIPTION OF PLATE XXIII

INSECTS —Fig 1 Oreina Ligea (the Arran brown Butterfly) female 2 The male 3 Showing the under side 4 The Caterpillar

„ Fig 5 Oreina Blandina (the Scotch Argus B) male 6 The female 7 The under side of the English variety, male 9 The under side of the Scotch variety, male 8 The under side of the English variety, female 10 The under side of Scotch variety female

PLANTS.—Fig 11 Poa Glauca (Glaucous Meadow-grass)

„ Fig 12 Poa alpina (Alpine Meadow grass)

I have been enabled from the fine cabinet of Mr Stephens to enrich the present plate with a variety of specimens illustrative of the distinctions of the Scotch and English varieties of O Blandina It will be seen that in the English specimens, the light band on the hind wing is ashy grey, broad in the female, and narrow in the male, and that in the female the grey band is repeated towards the base of the wing In the Scotch specimens, the light band on the hind wing is yellowish brown in the female, and in the male almost invisible from being scarcely lighter than the ground colour, whilst the secondary band is nearly obliterated in both sexes I regret much not being able to figure the caterpillar of this species, but hope some of our subscribers will enable us to supply the deficiency before the completion of the work H N H

---

* From the Greek 'Ορεινὸς *montosus*, from the species generally frequenting mountain districts

No. IX.] [Price 2s. 6d.

BRITISH INSECTS AND THEIR TRANSFORMATIONS.

BRITISH BUTTERFLIES

AND

Their Transformations.

ARRANGED AND ILLUSTRATED IN A SERIES OF PLATES

BY H. N. HUMPHREYS, ESQ.

WITH CHARACTERS AND DESCRIPTIONS

BY J. O. WESTWOOD, ESQ., F.L.S.,

SEC. OF THE ENTOMOLOGICAL SOCIETY, ETC. ETC.

LONDON:

WILLIAM SMITH, 113, FLEET STREET.

MDCCCXLI.

BRADBURY AND EVANS, PRINTERS, WHITEFRIARS.

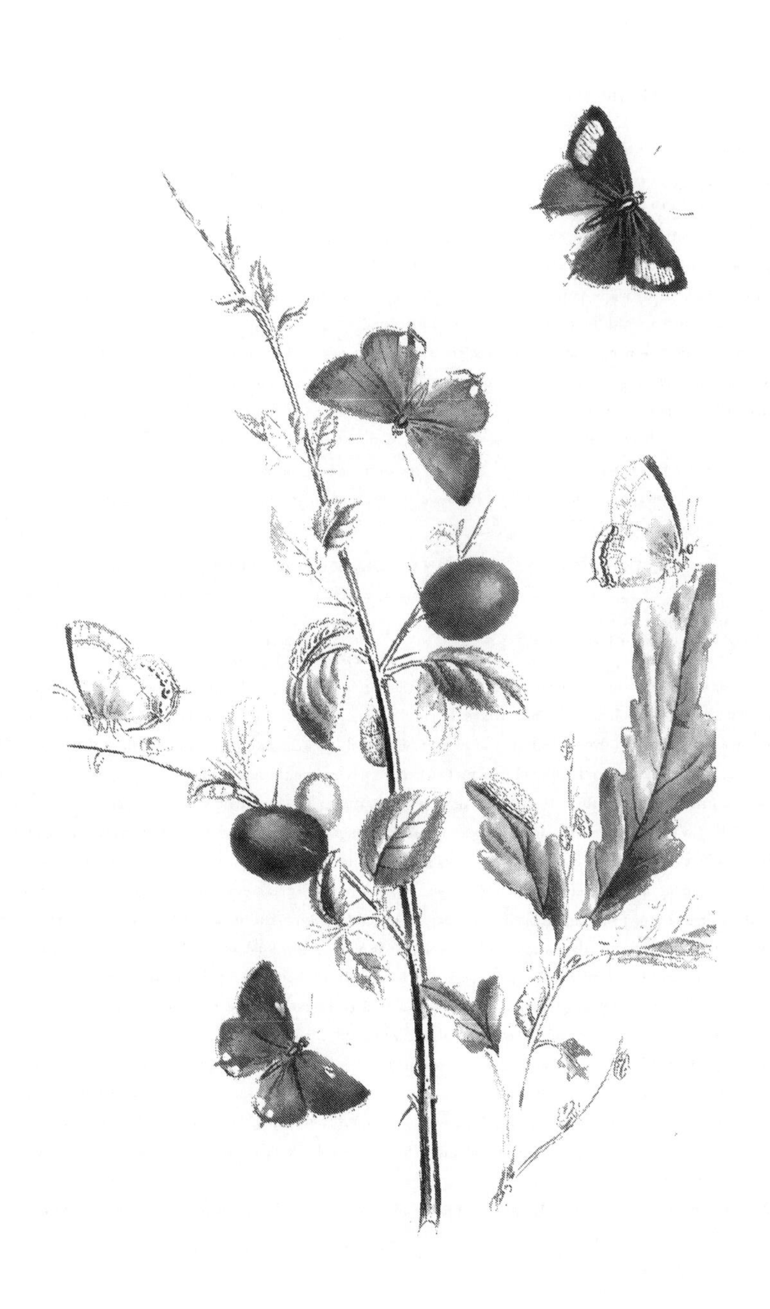

SPECIES 1.—OREINA LIGEA. THE ARRAN BROWN BUTTERFLY.

Plate xxiii. fig. 1—4.

SYNONYMES.—*Papilio Ligea*, Linnæus; Sowerby, Brit. Miscell. pl. 2.
Hipparchia Ligea, Ochsenheimer; Stephens, Illust. Haust. vol. 1. pl. 6, fig. 1, 2, 3; Duncan Brit. Butt. pl. 25, fig. 1.
Papilio Alexis, Esper.
Papilio Philomela, Hubner Pap. f. ...
Erebia Ligea, Dalman, Boisduval.
Erynnis Ligea, Hubner (Verz. bek. Schmett.)

This rare butterfly measures from $1\frac{5}{6}$ to 2 inches in the expanse of its wings, which are of a dark rich brown, all the wings having a broad oblong patch of red near the outer margin, within which, on the fore wings are four eyes, black (with white pupils in the females), the two nearest the apex confluent; and in the hind wings three ocelli, which are also blind in the males. On the under side the wings are of a paler brown, and the band is brighter in the fore wings, but almost obsolete in the hind ones, which are ornamented beyond the middle with an abbreviated and irregular white band, between which and the hind margin are three black ocelli, with white pupils, each surrounded by a red ring. The fringe of all the wings is alternately brown and white.

Taken in the Isle of Arran by the late Sir Patrick Walker and Alexander Mackay, Esq. in July or August. It is described as occurring in France and Sweden, and as appearing in meadows and open spaces of woods.

The caterpillar is green, with a dusky line down the back, with several white lines along the sides.

SPECIES 2.—OREINA BLANDINA. THE SCOTCH ARGUS BUTTERFLY.

Plate xxiii. fig. 5—10.

SYNONYMES.—*Papilio Blandina*, Fabricius; Sowerby, British Miscell. 1, pl. 7?; Donovan vol. 12, pl. 420.
Hipparchia Blandina, Ochsenheimer; Stephens; Curtis; Duncan Brit. Butt. pl. 25, fig. 2.
Epigea Philomela, Hubner.

This species varies in the expanse of its wings from $1\frac{5}{6}$ to 2 inches. The upper side of all the wings is of a dark uniform brown colour, the fore wings having a dark orange patch near the apex, the lower part being narrower than the upper, from which it is ordinarily separated by a constriction in the middle of the patch. The upper part of this patch bears a united pair of black eyes, having white pupils, and the lower part has a single eye similar in colour but smaller, which is occasionally obliterated. The hind wings are ornamented with a curved bar (or rather a united series of round marks), of obscure orange near the hinder margin, in which are generally three small black eyes with white pupils, and a black dot in the outer part of the bar. Specimens occasionally occur with as many as five ocelli on the fore wings. Others have only two ocelli in the hind wings. The fringe of all the wings is brownish, but paler and interrupted in the females. On the under side the fore wings are of a somewhat redder brown with an orange bar ocellated as above, the hind wings have the base greyish brown succeeded by a broad irregular red-brown bar, which extends beyond the middle of the wings, this is again succeeded by a rather narrower greyish bar, in which are three minute rudimental ocelli, and the margin of the wing is brown. The colouring of these bars varies considerably, not only according to the localities where the specimens are taken, but also in the sexes, as represented in the upper series of figures in Plate 23. Ordinarily, however, the specimens taken in Scotland have the bars more indistinctly marked. There are numerous Continental species closely allied to this insect, but the descriptions and figures of Ernst, Hubner, &c. of these insects are not sufficiently accurate to enable a correct judgment to be formed whether the English specimens be specifically distinct from the Scotch ones.

Messrs. Stephens and Curtis only mention the island of Arran and Castle Eden Dean as the localities for this species, but Mr Duncan adds, that "it occurs in some plenty over a district of considerable extent in Dumfries-shire near Minto in Roxburghshire, occasionally near Edinburgh, and probably in most of the southern counties of Scotland Mr Wailes informs us that it exists in profusion in one or two places in the magnesian limestone district not far from Newcastle The caterpillar is light green with brown and white longitudinal stripes, head reddish The egg is ribbed, and of a whitish colour speckled with brown

SPECIES 8—ORFINA CASSIOPE THE SMALL RINGLET BUTTERFLY

Plate xxii figs 9—10

SYNONYMS —*Papilio Cassiope*, Fabricius
Hipparchia Cassiope, Ochsenheimer, Stephens, Illustr Haust I pl 8 figs 1, 2, 3, Curtis, Duncan, Brit Butt pl 24, fig 5
Melampias Cassiope, Hubner
Papilio Æthiops minor, Villers
Papilio Melampus, Esper
Var —*Papilio Mnemon*, Haworth Ent Trans I 332

This species differs from the two preceding species of this group in having the wings much more elongated the hind pair being also entire and not denticulated The fore wings generally measure $1\frac{1}{4}$ or $1\frac{1}{2}$ inches in expanse The wings above are of a brown colour with a silky gloss, the fore wings having a red bar near the extremity interrupted by the veins, and not extending to the margins of the wings in this bar are generally four small black dots with obscure pupils but specimens occur with only three, two or even no ocelli, whilst in others the bar itself is reduced to a few red spots The hind wings have also a red bar near the extremity, bearing three similar eyelets On the under side the fore wings are redder brown, with the red band marked with four black spots, whilst the hind ones are ashy or coppery-brown, with three black spots, each surrounded by a slender red ring Variations occur in the number and size of the spots as well as of the band

The mountainous districts of Cumberland and Westmoreland are the only localities yet mentioned for this small species, upon which Mr Curtis makes the following observations —"The males in forward seasons have appeared as early as the 11th of June, but last year (1829), when Mr Dale and myself visited Ambleside, they were later, the first being taken the 18th of June, and they did not become plentiful till the 25th They are found amongst the coarse grass that covers considerable spaces abounding with springs on the sides of mountains They only fly when the sun shines, and their flight is neither swift nor continued, for they frequently alight amongst the grass and falling down to the roots their sombre colour perfectly conceals them The females are later, and have been taken even in August We found the males on Red Skrees, a mountain near Ambleside, and Mr Marshall took them at Gable-hill and Sty-head, between Wastwater and Borrowdale"

HIPPARCHIA MNESTRA of Ochsenheimer was introduced into the list of English species without authority in the 2nd Edition of the Butterfly Collector's Vade-Mecum, on the examination of a variety of Orfina Cassiope in the British Museum, in which the fascia on the fore wings has only two eyelet spots Mr Stephens, however, corrected the error in his Illustrations (Haustell vol i p 63), but Mr Curtis has subsequently given Mnestra as a British species in the second edition of his Guide, but accompanied by a mark of interrogation That gentleman has, however, recently observed to me in a note, "You will observe that Mnestra Hub ? is *queried* and it may be only the female of Cassiope" The true Mnestra, as carefully figured by Boisduval in his Icones Historiques des Lepidopteres d'Europe, vol i pl 35 figs 1—4, has the disc of the under side of the fore wings in both sexes rich red brown—the four wings are also "proportionnellement assez courtes arrondies" It is found in various parts of Switzerland especially near the Great and Little St Bernard The males have the red band of the fore wings unspotted, and in the females it has two eyes on each side

FAMILY III

ERYCINIDÆ

This family of butterflies is distinguished by the males having only four ambulatory feet, whilst the females have six, or in other words, the fore legs of the males are rudimental whilst they are perfect in the females, the anal edge of the hind wings is but slightly prominent, the discoidal cell is either open, or closed either entirely or partially by a false nervure The claws of the tarsi are minute and scarcely perceptible The caterpillars are very short, pubescent or hairy, and the chrysalis is short and contracted

The insects of this family are of small size, and almost exclusively confined to South America They are often very brilliant and varied in their colours, their wings being mostly marked with spots They, however, exhibit a certain appearance of weakness in their formation quite unlike that of Nymphalidæ M Lacordaire, nevertheless informs us that the flight of the South American species is very rapid and that the majority rest with their wings extended on the under side of the leaves The only British species which belongs to this family (forming, indeed, an aberrant group therein) is the small fritillary known to collectors under the name of the Duke of Burgundy, which differs from all the Nymphalidæ (in which family it has been arranged by Stephens Curtis, &c,) in several important respects especially the perfect structure of the fore legs in the female the minute simple ungues, the posterior tibiæ destitute of spurs, the onisciform larva, and the girt chrysalis In its general appearance, however, as well as in its colours and markings, it bears a more immediate resemblance to the small fritillaries of the genus Melitæa, but the relation is one of analogy and not of affinity, the general appearance alone constituting the resemblance, whilst in its more important structural characteristics it possesses no real relation with the Melitææ How far it would be advisable (as has been suggested to me by Mr Stephens) to invert the arrangement of the genera constituting the family Nymphalidæ in order to bring the Hipparchides into conjunction with the Pierides, and thus terminate the family with the Melitææ (which would be thus brought into connexion with the present family), can only be determined when a complete analysis has been made of the family Nymphalidæ— a task which it is to be feared no lepidopterist is yet able to accomplish Such a step would, moreover, completely overturn the arrangement of Dr Horsfield, and the *natural* transition of the genera which he laboured to propound

DESCRIPTION OF PLATE XXIV

INSECTS —Fig 1 Hamearis Lucina (the Duke of Burgundy Fritillary B) 2 Showing the under side 3 The Caterpillar 4 The Chrysalis

" Fig 5 Thecla Quercus (the purple hair streak B), female 6 The male 7 A common variety of the male, more dusky in tint 8 Showing the under side 9 The Caterpillar 10 The Chrysalis

PLANTS —Fig 11 Primula vulgaris (the common Primrose) 12 Quercus pedunculata (a variety of the common Oak)

I have described fig 1 as the female T Quercus, a fact said to be ascertained by a careful dissection But as in every other instance among diurnal Lepidoptera, the fine *metallic* tints, similar to the brilliant purple marks on the fore wings of this insect, are only displayed by the male— as in the purple Emperor, all the genus Lycæna, &c , &c —this assertion is against general analogy Both insects are from specimens in the British Museum, and the larvæ are from Hubner H N H

The only British genus is the following —

HAMEARIS, Hubner. (NEMEOBIUS, Stephens.)

This very interesting genus is distinguished from all the preceding by the small size of the insects, as well as from all the Nymphalidæ by the nature of its transformations. The head is rather small, with the eyes hairy, and the short palpi not extending beyond the tuft of hairs on the forehead; they are slender and three-jointed, the second joint long, and the third small and sub-globose. The antennæ are slender and terminated by an abrupt compressed club. The fore wings are short and somewhat triangular, the front margin being nearly straight; the hind wings are rounded and denticulated. The veining of the wings is peculiar and has not been before described. In the fore wings the veins are arranged as in Argynnis, the postcostal vein superiorly emitting two branches before reaching the transverse vein which closes the discoidal cell. A third branch is emitted at the place of the junction of the transverse with the postcostal veins; this third branch superiorly emits two branchlets; the postcostal vein after emitting this third branch is simple, and extends to below the tip of the wing. In the hind wings the vein which corresponds with the postcostal one emits an outward branch (which extends to the outer angle of the wing), at a considerable distance *below* the place from whence the interior branch of this postcostal vein is emitted.* Whereas in the other fritillaries the outward branch originates considerably nearer the base of the wing than the inward branch. The fore legs are short, imperfect, and thickly clothed with hair in the males, the tarsal portion being destitute of articulations. In the female the fore legs, on the contrary, are as perfect as the others, and clothed with short scales, the tarsi being as long as the tibia, composed of five distinct joints, terminated by minute simple claws and pulvilli. The tibiæ of the hind feet are destitute of spurs.

Mr. Stephens, with philosophical tact, pointed out the propriety not only of separating this genus from the other fritillaries, but also noticed its variance in several respects with the characters of the family Nymphalidæ, in which he placed it, such as the simplicity of the claws and posterior tibiæ. He was, however, unacquainted with the caterpillar and chrysalis states. These have, however, since been figured by Mr. Curtis (copied from Hübner) and fully prove the correctness of Mr. Stephens' suggestion of the distinction between this genus and the other Nymphalidæ, the caterpillar being oniscιform and the chrysalis not only attached by the tail, but also fastened by a girth across the middle of the body.

The only British species is

HAMEARIS LUCINA. THE DUKE OF BURGUNDY FRITILLARY.

Plate xxiv. fig. 1—4.

Synonyms.—*Papilio Lucina*, Linnæus. Lewin, Br. Pap. pl. 15. Donovan Brit. Ins. vol. 7, pl. 212. f. 2. Harris, Aurelian, pl. 27, fig. n—o. Haworth, (not of Wilkes, pl. 111, which represents Artemis.) *Nemeobius Lucina*, Stephens, Horsfield, Duncan Brit. Butt. pl. 12, fig. 1. Boisduval, Hist. Nat. Lep. vol. 1, pl. 2 B, fig. 8. Westwood, Introd. to Mod. Classif. of Ins. vol. 2, p. 357, fig. 99, 12—14. *Hamearis Lucina*, Hubner. Curtis, Brit. Ent. pl. 516. *Melitæa Lucina*, Ochsenheimer. Leach. Jermyn.

This pretty little butterfly varies in the expanse of its wings from 1 to 1¼ inches. On the upper side the ground colour of the wings is brown, ornamented with numerous orange-coloured marks and spots, forming in the fore-wings three transverse series. The one nearest the base of the wings is most irregular, forming two strong curves; in the second, the spots are of unequal size, the two nearest the costa being very

* I have only found this peculiarity in several species of Erycinidæ (see Boisduval, Hist. Nat. Lep. Pl. 20, 21.)

minute, and the two following the largest; the terminal row, consists either of a series of oval spots, with a black dot in the centre of each, or orange crescents. The cilia is alternately black and white; the hind wings are brown with about four small irregular pale spots, and a submarginal row of spots or lunules as in the fore wings. Beneath, the wings are much more varied in their colours, the ground colour being much paler; the orange spots again appear, but are separated from each other by black marks towards the hind margin of the fore wings; the hind wings are ornamented towards the base with an abbreviated bar, formed of four or six white oblong spots; this is succeeded by an orange bar, which is followed by an irregular white bar, formed of nine unequal-sized white patches. Along the margin is a row of fulvous ocelli, with a black dot in the middle of each, edged behind with white. The ground colour of the wings of the female is black instead of brown, with the markings larger and brighter coloured; the latter in the males are occasionally almost obliterated, except the marginal ones.

The following account of the preparatory states of this interesting insect, is quoted by Mr. Curtis, from Hubner's valuable work on the European Lepidoptera:—"The eggs are found solitary, or in pairs, on the under surface of the leaves of *Primula veris* and *elatior* at the beginning of summer; they are almost globular, smooth, shining, and pale yellowish-green. The caterpillar feeds on the leaves; its head is roundish, heart-shaped, smooth, shining, and bright ferruginous, black only on the mouth and about the eyes; its body is almost oval, but long, depressed, and set with rows of bristly warts; the other parts are set with feathery hairs; on the back, at least from the fourth joint to the tail, there is a black dot on each joint, and on the sides similar but less distinct spots; the colour is pale olive orange; its feet are rusty brown, the spiraculæ black; claws and belly whitish. It moves very slowly, rolls itself up when disturbed, and remains in that state a long time. Soon after the middle of summer it becomes a pupa, not only fastening its body by the apex, but also by spinning a cord across its middle. In this state it remains until the end of the following spring. Hubner, who reared it from the egg, says also that the caterpillar throws off five skins before it becomes a pupa, and its appearance, at different ages, varies considerably. The larva represented, (and copied in our plate xxiv. fig. 3,) he found on a *primula* in his own garden."

Coombe Wood; Darenth, Kent; Boxhill; Dulwich; the New Forest; and in Dorsetshire, Berkshire and Northamptonshire, are given for the localities of this rather uncommon species; and Mr. Duncan adds that it has been also taken as far north as the neighbourhood of Carlisle, by Mr. Heysham.

FAMILY IV.

LYCÆNIDÆ, LEACH.

The present family, (corresponding with the Polyommatidæ of Swainson and the Vermiform Stirps of the butterflies of Dr. Horsfield), comprises a numerous assemblage of small and weak, but beautiful creatures, distinguished by the minute size of the tarsal claws, the apparent* identity in the fore tarsi of both sexes, the fore

---

* The fore tarsi have been described by Messrs. Curtis, Stephens, &c., as identical in both sexes; but in examining the Indian Theclæ [illegible] I discovered that the tarsus of the males consists of a long simple joint; and I subsequently found the same to be the case in Polyommatus Corydon.

legs being fit for walking, the hind tibiæ with only one pair of spurs, the antennæ not distinctly hooked at the tip, the last joint of the palpi is small and naked, the anal edge of the hind wings slightly embraces the abdomen, the discoidal cell of the hind wings is apparently closed by a slender vein. The caterpillars bear a very considerable resemblance to wood-lice, the head being retractile, and the feet very minute, the body is oval and depressed: the chrysalis is short, obtuse at each end, and girt round the middle as well as attached by the tail.

The family comprises several distinct groups, namely, such as are known to collectors under the names of blues, coppers, and hair-streak butterflies, respectively distinguished in our indigenous species by their varied tints of blue, fiery-red, or dusky, with slender lines on the under side of the wings. In many of the exotic species, however, these colours run beyond the limits of their respective groups, thus forming a series. The majority have the entire under surface of the wing, or at least the anal angle, ornamented with beautiful eye-like spots of various colours. Some of the exotic species are amongst the most lovely of the butterfly tribes. In many of these, the hind wings are produced into very long tails. Their flight is varied, some delighting to sail over the tops of oaks and other trees, on which they have passed their preparatory states, whilst others are feeble and slow in their motions, flying over low grass and herbage.

Dr. Horsfield has investigated the transformations of many of these insects, in his *Lepidoptera Javanica*, the larvæ of which vary very considerably in their form, some exhibiting a much slighter resemblance to wood-lice than others; some are very rough on the upper surface of the body; and that of Thecla Xenophon, a Javanese species, has several rows of fascicles of hairs. They have hitherto been observed to feed only upon the leaves of different plants and trees in the larva state; but a beautiful Indian species (Thecla Isocrates) resides within the fruit of the pomegranate—several (seven or eight) being found within one fruit, in which, after consuming the interior, they assume the pupa state, having first eaten as many holes as there are insects through the rind of the fruit, and carefully attached its foot-stalk to the branch by a coating of silk, in order to prevent its falling. (Westwood, in Trans. Entomological Society, vol. ii. pl. 1.)

I have already (in p. 35) referred to the arrangements which have been proposed by various authors, in order to bring this family, with its girt chrysalis, into connexion with the Papilionidæ, in which the chrysalis is also girt. We are not sufficiently advanced in the knowledge of exotic Lepidoptera to determine the soundness of these different views.

GENUS XXI.

THECLA, Fabricius.

This genus, as characterised by our English species, is at once known from the two following by the very gradually formed club to the antennæ, the short terminal joint of the palpi, the hairy eyes*, the short triangular fore wings and hind wings generally furnished with a short tail and strongly scalloped near the anal angles, and by the under surface of the wings being generally ornamented with one or two delicate lines of a pale colour on a

* Mr. Curtis, Brit. Ent. pl. 264, represents the eyes of this genus as naked, although he describes them correctly.

dark ground This last-mentioned peculiarity has led to the insects of this genus being named by collectors hair-streak butterflies "

No genus has ever more strongly proved the great advantages to be derived from a minute, analytical examination of the different organs of insects than the present For want of such an examination in the structure of the feet which in this group affords sexual differences, the males of the beautiful species represented in our 24th Plate are described in all previous works on English butterflies as the females, and vice-versâ Even by Messrs Curtis and Stephens, who introduce the structure of the fore feet into the characters of the genus, they are described as "alike in both sexes" For our knowledge of the peculiarities existing in these organs in the different sexes we are indebted to Dr Horsfield, who was accordingly enabled to determine the sexes of the species with precision The male anterior tarsus consists of a single long joint, which is as long as the entire *articulated* tarsus of the female, and when covered with scales might easily be regarded as similarly articulated The intermediate thigh is also furnished with a remarkable tooth near the extremity and there is a corresponding notch in the tibia

In preceding pages (9 and 23), I have described the two different plans adopted by elongated Caterpillars in order to effect their transformation beneath a girth across the middle of the body The caterpillars of the present family, by their short wood-louse form, at first sight present far more apparent obstacles to the accomplishment of such a proceeding We have only to fancy to ourselves a very stout little man with his arms bound to his sides, and his legs tied together laid upon his belly on the ground and whilst in that situation compelled to attach a rope on each side of his body, carrying it across his back and then undressing himself Reaumur has, however so carefully and circumstantially described the proceedings of a species of this genus*, that we are enabled to bring the whole process before our mind's eye When the period for the transformation of the insect is arrived, the caterpillar attaches itself by the tail and shortening the fore part of the body by contracting the segments considerably, it emits an arched thread from the extended spinneret of its mouth, which it attaches at one side of the head it then carries the thread to the other side, having the instinct to extend it to a fit length in order that when the process is complete the body may neither be too much tightened nor too loose for support After attaching a number of threads by passing the head backwards and forwards, emitting a continuous thread from the mouth during the process, a skein of between 50 and 60 threads (as Reaumur supposes) is formed, of a fit length, attached at each end at the sides of the head, which, as already stated, is drawn considerably backwards, owing to the great contraction of the anterior segments of the body In this consists the chief difference between the proceedings of these insects and those of the swallow-tailed and cabbage butterflies, described in our previous pages, the head of the caterpillars of those species, owing to the slender form of the body, being thrown over the back, or greatly elevated During the process of spinning this skein of thread, the caterpillar has contrived to insinuate its head beneath them, so that the skein rests upon the scaly back of the head, and when the skein is completed it gradually pushes the front of its body beneath the skein, pressing its body down as closely as possible, until it contrives that the middle of the body shall be girt by the skein The difficulty is apparently increased by the very great delicacy of the threads, and by the body of the larva being clothed with strong short bristles When completed, the skin of the caterpillar bursts and the pupa appears, the process of getting rid of the old skin being similar to that adopted by the other girt caterpillars of the butterflies mentioned above

* Linnæus gives Reaumur's insect as his Papilio Pruni, but this can scarcely be, for Reaumur's description does not accord with that species and Reaumur found his larva on the elm

The caterpillars of this genus are not found on herbaceous plants, but frequent trees and shrubs, over which the perfect insects fly. The species are very numerous, although we possess but very few, and even amongst those some very peculiar distinctions occur. Thus, in Thecla Quercus (contrary to the general rules of insect colouring), the female puts on the "imperial purple;" the fulvous patch in the fore wing of the female of Thecla Betulæ is found in the females of some exotic species, the males of which are adorned with the purple tint, whereas the males of Thecla Betulæ are obscurely coloured. The males of Thecla Rubi, Spini and W-album are distinguished by having a small ovate glabrous patch at the extremity of the discoidal cell of the fore wings on the upper side*. Th. Rubi differs from all the rest in not having the hind wings tailed, and by the underside of the wings being neither marked with the slender pale hair-streak, nor by an ocellus at the anal angle.

Dr. Horsfield, who has described twenty-six species of these insects, found in Java, has divided the genus into two subgenera, Thecla proper and Amblypodia. Unfortunately, however, he did not investigate the peculiarities of the veins of the wings of his subgenera and sections, a character which the reader need scarcely be reminded has been already shown, in the pages of this work, to be of primary importance in determining natural groups. Although previously so greatly neglected, this character supplies the means of dividing our English species into two primary groups†, which ought, perhaps, consistently to be considered as distinct subgenera, supported as they are by some other characters. It is, however, perhaps more advisable (until the exotic series of species is carefully investigated), to leave the genus entire, indicating the groups into which the British species are divisible. This plan is also adopted, because the investigation of the peculiarities in the veining of the wings is attended with great difficulty, the scales having to be carefully removed from the surface of the wings. It is certainly remarkable that we should find, in species so closely allied together as all the British Theclæ are, such a variation in the veins, more especially as we have seen that nearly all the Nymphalidæ, varying as they do, so greatly in their preparatory as well as perfect forms, exhibit an identity in the arrangement of these veins—but Nature, in every extensive group, shows us the impropriety of trusting to a single character, which, in some tribes, may be most important and constant, whereas, in others, it may become variable and of secondary importance.

Retaining Thecla Betulæ as the true Fabrician type of the genus, I divide it in the following manner.

I. Those in which the postcostal vein of the fore wings emits two branches before its union with the ordinary transverse vein, and a third branch beyond its union therewith; this third branch sending forth a superior branchlet. Males without a patch at the extremity of the discoidal cell of the fore wings; antennæ with the club very gradually formed.

Sp. 1. Thecla Betulæ. 2. Thecla Quercus.

II. Postcostal vein of the fore wings emitting, in both sexes, three simple branches before and none after its union with the ordinary transverse vein. Males with a thickened patch at the extremity of the discoidal cell of the fore wings; antennæ with the club more suddenly formed.

A, with the hind wings tailed. 3. Thecla Pruni. 4. Thecla W-Album. 5. Thecla Spini. 6. Thecla Ilicis.

B, with the hind wings not tailed. 7. Thecla Rubi.

---

* This is produced by the dilatation of the base of the 2nd and 3rd branches of the postcostal vein.

† Hübner unites all the Theclæ into one family (Fam. C. Armati), of his Adolescentes, separating Rubi under the generic name of Lycus, and Quercus under that of Bithys; the remainder forming his Strymon. Ochsenheimer gives our Thecla as the third family of Lycæna, separating however, Rubi from the rest, and uniting it with the second family containing the Coppers.

ILLUSTRATED EDITION OF FROISSART.

In two thick Volumes, price 36s

SIR JOHN FROISSART'S CHRONICLES OF ENGLAND, FRANCE, SPAIN, &c.

THIS Edition is printed from the Translation of the late THOMAS JOHNES, Esq, and collated throughout with that of LORD BERNERS numerous additional Notes are given, and the whole embellished with *One Hundred and Twenty Engravings on Wood*, illustrating the Costume and Manners of the period chiefly taken from the illuminated MS copies of the Author in the British Museum, and elsewhere

COMPANION TO FROISSART.

In two Volumes, price 30s

THE CHRONICLES OF MONSTRELET.

WITH NOTES AND WOODCUTS, UNIFORM WITH THE ABOVE EDITION OF FROISSART

ILLUSTRATED EDITION OF "MARMION"

In demy 8vo, price 16s cloth, 21s morocco elegant,

M A R M I O N.

A Poem.

BY SIR WALTER SCOTT

WITH FIFTY BEAUTIFUL WOOD ENGRAVINGS

ILLUSTRATED WITH FIFTY-ONE PORTRAITS

BURNET'S HISTORY OF HIS OWN TIMES.

In Two Volumes, super-royal 8vo, cloth lettered, price 2*l* 2*s*, or half bound in morocco, 2*l* 12*s* 6*d*

ILLUSTRATED WITH FIFTY-SIX PORTRAITS

CLARENDON'S HISTORY OF THE REBELLION.

In Two Volumes, imperial 8vo, price 2*l* 10*s* cloth lettered

In one Volume, foolscap 8vo, price 7s. in cloth,

TREATISE ON THE INSECTS INJURIOUS TO THE GARDENER, FORESTER, AND FARMER.

TRANSLATED FROM THE GERMAN OF M KOLLAR, AND ILLUSTRATED WITH ENGRAVINGS

BY J AND M LOUDON

WITH NOTES BY J O WESTWOOD, ESQ

"We heartily recommend this treatise to the attention of every one who possesses a garden, or other ground, as we are confident that no one taking an interest in rural affairs can read it without reaping both pleasure and profit from its perusal"—*Literary Gazette*

"We have always wondered that, in a country like this, where the pursuits of agriculture and horticulture are so universal and important entomologists should never have bethought them of writing a book of this description It is, therefore, with great satisfaction, that we announce the appearance of the present translation of a work which goes far to supply the deficiency we have spoken of"—*Athenæum*

"From the very neat and cheap manner in which the volume is got up, we trust it will become a favourite, not only with the entomologist, but with every lover of agriculture arboriculture, and horticulture"—*Mag Nat Hist*, Feb 1840

(*To be completed in about Sixteen Numbers.*)

No. X.] [PRICE 2*s.* 6*d.*

BRITISH INSECTS AND THEIR TRANSFORMATIONS.

BRITISH BUTTERFLIES

AND

Their Transformations

ARRANGED AND ILLUSTRATED IN A SERIES OF PLATES

BY H. N. HUMPHREYS, ESQ.

WITH CHARACTERS AND DESCRIPTIONS

BY J. O. WESTWOOD, ESQ., F.L.S.,

SEC. OF THE ENTOMOLOGICAL SOCIETY, ETC. ETC.

LONDON:
WILLIAM SMITH, 113, FLEET STREET.

MDCCCXLI.

BRADBURY AND EVANS, PRINTERS, WHITEFRIARS.

SMITH'S STANDARD LIBRARY.

In medium 8vo, uniform with Byron's Works, &c

Works already Published

ROKEBY BY SIR WALTER SCOTT 1s. 2d

THE LADY OF THE LAKE 1s
THE LAY OF THE LAST MINSTREL 1s
MARMION 1s 2d
THE VISION OF DON RODERICK BALLADS AND LYRICAL PIECES By SIR WALTER SCOTT 1s
THE BOROUGH By the REV G CRABBE 1s 4d
THE MUTINY OF THE BOUNTY 1s 4d
THE POETICAL WORKS OF H KIRKE WHITE 1s
THE POETICAL WORKS OF ROBERT BURNS 2s 6d
PAUL AND VIRGINIA, THE INDIAN COTTAGE, and ELIZABETH 1s 6d
MEMOIRS OF THE LIFE OF COLONEL HUTCHINSON, GOVERNOR OF NOTTINGHAM CASTLE during the Civil War By his Widow, Mrs LUCY HUTCHINSON 2s 6d
THOMSON'S SEASONS, AND CASTLE OF INDOLENCE 1s
LOCKE ON THE REASONABLENESS OF CHRISTIANITY 1s
GOLDSMITH'S POEMS AND PLAYS 1s 6d
——— VICAR OF WAKEFIELD 1s
——— CITIZEN OF THE WORLD 2s 6d
——— ESSAYS, &c 2s

, These four Numbers form the Miscellaneous Works of Oliver Goldsmith, and may be had bound together in One Volume, cloth-lettered, price 8s

KNICKERBOCKER'S HISTORY OF NEW YORK 2s 3d
NATURE AND ART By MRS INCHBALD. 10d
A SIMPLE STORY By MRS INCHBALD 2s
ANSON'S VOYAGE ROUND THE WORLD 2s 6d
THE LIFE OF BENVENUTO CELLINI 3s
SCHILLER'S TRAGEDIES THE PICCOLOMINI, and THE DEATH OF WALLENSTEIN 1s 8d
THE POETICAL WORKS OF GRAY & COLLINS 10d
HOME By MISS SEDGWICK 9d
THE LINWOODS By MISS SEDGWICK 2s 8d
THE LIFE AND OPINIONS OF TRISTRAM SHANDY By LAURENCE STERNE 3s
INCIDENTS OF TRAVEL IN EGYPT, ARABIA PETRÆA, AND THE HOLY LAND By J L STEPHENS, Esq 2s 6d
INCIDENTS OF TRAVEL IN THE RUSSIAN AND TURKISH EMPIRES By J L STEPHENS, Esq 2s 6d
BEATTIE'S POETICAL WORKS, and BLAIR'S GRAVE 1s
THE LIFE AND ADVENTURES OF PETER WILKINS, A CORNISH MAN 2s 4d
UNDINE A MINIATURE ROMANCE Translated from the German by the REV THOMAS TRACY 9d
THE ILIAD OF HOMER Translated by ALEX POPE 3s
ROBIN HOOD, a Collection of all the ancient Poems, Songs, and Ballads now extant relative to that celebrated English Outlaw, to which are prefixed, Historical Anecdotes of his Life Carefully revised, from RITSON 2s 6d
THE LIVES OF DONNE, WOTTON, HOOKER, HERBERT, AND SANDERSON Written by IZAAK WALTON 2s 6d
THE LIFE OF PETRARCH By MRS DOBSON 3s
MILTON'S PARADISE LOST 1s 10d
——— PARADISE REGAINED, and MISCELLANEOUS POEMS 2s
RASSELAS By DR JOHNSON 9d
ESSAYS ON TASTE By the Rev ARCHIBALD ALISON, LL B 2s 6d
THE POETICAL WORKS OF JOHN KEATS 2s
THOMSON'S POEMS AND PLAYS, COMPLETE 5s
PICCIOLA A Tale from the French 1s 1d
THE POETICAL WORKS OF DR YOUNG, COMPLETE 5s

FOUR VOLUMES ARE NOW ARRANGED ON THE FOLLOWING PLAN —

ONE VOLUME OF "POETRY,"

CONTAINING

SCOTT'S LAY OF THE LAST MINSTREL
SCOTT'S LADY OF THE LAKE
SCOTT'S MARMION
CRABBE'S BOROUGH
THOMSON'S POETICAL WORKS
KIRKE WHITE'S POETICAL WORKS.
BURNS'S POETICAL WORKS

A SECOND VOLUME OF "POETRY,"

CONTAINING

MILTON'S POETICAL WORKS
BEATTIE'S POETICAL WORKS
BLAIR'S POETICAL WORKS
GRAY'S POETICAL WORKS
COLLINS'S POETICAL WORKS
KEATS'S POETICAL WORKS
GOLDSMITH'S POETICAL WORKS

ONE VOLUME OF "FICTION,"

CONTAINING

NATURE AND ART By MRS INCHBALD
HOME By MISS SEDGWICK
KNICKERBOCKER'S HISTORY OF NEW YORK By WASHINGTON IRVING
PAUL AND VIRGINIA By ST PIERRE
THE INDIAN COTTAGE By ST PIERRE
ELIZABETH By MADAME COTTIN
THE VICAR OF WAKEFIELD By OLIVER GOLDSMITH
TRISTRAM SHANDY By LAURENCE STERNE

ONE VOLUME OF "VOYAGES AND TRAVELS,"

CONTAINING

ANSON'S VOYAGE ROUND THE WORLD
BLIGH'S MUTINY OF THE BOUNTY
STEPHENS'S TRAVELS IN EGYPT, ARABIA PETRÆA, AND THE HOLY LAND
STEPHENS'S TRAVELS IN GREECE, TURKEY, RUSSIA, AND POLAND

These Volumes may be had separately, very neatly bound in cloth, with the contents lettered on the back, price 10s 6d each

Those parties who have taken in the different works as they were published, and who wish to bind them according to the above arrangement, may be supplied with the title-pages gratis, through their booksellers, and with the cloth covers at 1s each

JOURNAL OF CIVILISATION.

On the Eighth of May, 1841, will be published, to be continued on each succeeding Saturday,

IN IMPERIAL 8VO, BEAUTIFULLY ILLUSTRATED WITH WOOD CUTS,

PRICE THREEPENCE,

THE JOURNAL OF CIVILISATION.

BLISHED UNDER THE SUPERINTENDENCE OF THE SOCIETY FOR THE ADVANCEMENT OF CIVILISATION

A SPIRIT of inquiry, observation, and research is one of the most prominent characteristics of the present age stead of a solitary traveller here and there, looking on man in s greatly diversified condition and then recounting his adven- es in his own circle, there is a multitude traversing the globe, ying into the state of its inhabitants, and laboriously describ- g it to others Never were our facilities for studying human ture so many, and yet, though daily on the increase, they are happily available only to a very limited extent

It is an indisputable fact, that many travellers record their scoveries in memoirs read before our learned societies, frag- ents of which only meet the eyes of the few in the pages of their Transactions," while others publish them in volumes so large d costly, as to render their use equally exclusive In some instances it is otherwise, but the purchase of what is reall desirable would amount annually to a considerable sum

The ignorance of our race which is thus sustained with th various evils springing from it, supplies a powerful motive for th diffusion of sound and valuable knowledge in another and a nove form, and this is enforced by important considerations Mar wherever he appears, has indefeasible claims to all the privilege of brotherhood A savage, he asks to be civilised, ignorant, t be taught, wretched, to be made happy, prostrate in abjec and torturing degradation, to be raised to the intellectual an moral dignity of a being created in "the image of God" No are cases wanting, in which distance, often reducing his voice to whisper, and even rendering it totally inaudible, is now abso lutely annihilated, and the connexion between us and him is a intimate as that which subsists between ourselves and the vas mass of human life amidst which we daily and hourly move The suppliant for aid, although adorned with other colours an speaking another language, dwells in *our* village, in *our* street at *our* door

For are not our countrymen, our kindred, increasing the colo nies of Britain, where there have long been other owners of th soil? And do not the aborigines of these climes urge on us a irresistible plea? Conquest, too, is extending the range o Britain's sceptre, and ought not the vanquished to feel that it power is peculiarly benign? There was always a bond betwee ourselves and the New Zealanders, or the people of Afghanista —the bond arising from oneness of blood—but recent events hav added largely to its fibres, and given it unprecedented tenacit and strength.

Heie then is the basis of oui intended publication its object to exhibit to all whom we can reach, man as he is, and the ieans iequired, and in action for the amelioration of his condi-on So far from confining oui view to scenes abroad, we shall faithfully portray scenes at home To others there may be a owerful attraction in a broken language and a copper-coloured kin, but with us interest as intense as theirs will be awakened hen the hue that meets us is our own, and the accents in which e are addressed are those of our mother-tongue. Man shall istain no less from us because his dwelling is distant, nor shall is demands be concealed or exaggerated because he is near.

On these grounds we ask for cordial and comprehensive sup-ort We cannot conceive of the individual for whom no pro-ision will be made, except, indeed, he be divested of the common elings of humanity, or ignorant of the English language We ave nothing to do with any political party The man of business nay, therefore, glean from our pages information as to the state f our Colonies and Eastern Dependencies, and descry Commerce s the invariable companion of Christianity and civilisation; the an of literature may find samples of modern travels either to atisfy or stimulate his thirst and the man of science may trace, n the fullest confidence, the progress and results of geographical iscovery All will be aided by Pictorial Illustration where erbal description fails, or by which it may be made more pleasing nd effective Maps of Countries, coasts, and towns and ngravings of localities peculiarly interesting, of the aspect and sages of the various tribes of men, and of the animal and egetable products which most nearly concern the human family vill be prepared in the highest style of accuracy and beauty

To execute the task proposed is, indeed, to provide for all. Even the mother may gather much for her child; the youth may peruse narratives more heart-thrilling than fiction ever penned; those who visit or reside in foreign lands may have the news of the world within their grasp; others who stay at home, may enjoy many results, with none of the pains of adventure; the rich may read what would not otherwise be observed, at least without offence; the poor may do so with manifest advantage.

On the zealous support and vigorous co-operation of all the friends of Christian missions we fully calculate. We adopt no exclusive creed; we yield to no sectarian bias. "Missionaries of every sect when among the heathen nations, have in every instance, without a single exception, given each other the right-hand of fellowship in cordial co-operation;*" one object constantly kept before us will, therefore, be to render their harmony, if possible, more influential. It is now an axiom that true religion is the great parent of civilisation; philanthropy claims the same lineage; and we, therefore, proceed to our projected work energetically and hopefully, impelled by the sincere desire greatly to diminish human woe and augment the means of human happiness.

* Quarterly Review.

LONDON:
PUBLISHED BY WILLIAM SMITH, 113, FLEET STREET;
AND SOLD BY
J. MENZIES, EDINBURGH; R. GRIFFIN & CO. AND D. BRYCE, GLASGOW; CURRY & CO., DUBLIN; SIMMS & DINHAM, C. AMBERY AND R. WHITMORE, MANCHESTER; G. PHILIP AND ARNOLD & SON, LIVERPOOL;
AND BY ALL BOOKSELLERS AND NEWSMEN IN THE KINGDOM.

[illegible] Printers, Whitefriars.

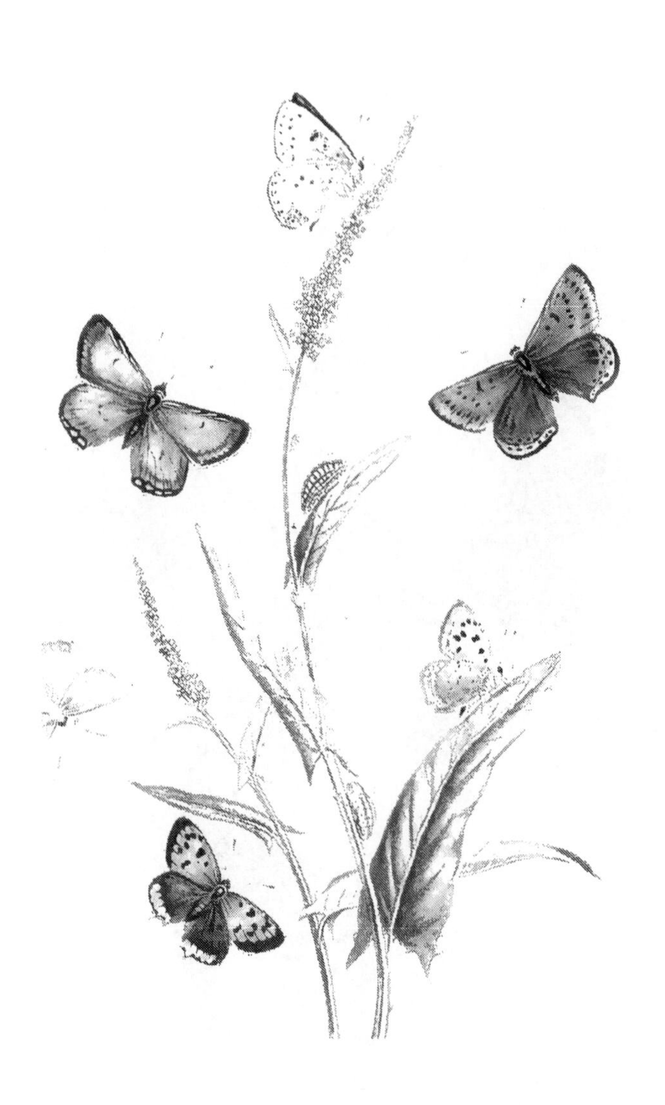

DESCRIPTION OF PLATE XXV

INSECTS.—Fig 1 Thecla Betulæ (the brown hair-streak B.), female 2 The male 3 Showing the under side 4 The Caterpillar 5 The Chrysalis

" Fig 6 Thecla Pruni (the black hair-streak B.), male 7 The female 8 Showing the under side 9 The Caterpillar 10 The Chrysalis

PLANTS.—Fig 11 Prunus domestica (the wild Plum) 12 Betula alba (the common Birch)

T Betulæ is from specimens in the British Museum. T Pruni from specimens in the fine collection of Mr Stephens. The larvæ of both are from Hubner. The under side, fig 3, is a female, the markings of the male are much less distinct and of a more general rust colour. H. N. H.

SPECIES 1.—THECLA BETULÆ. THE BROWN HAIR-STREAK.

Plate xxv fig 1—5

SYNONYMES.—*Papilio Betulæ*, Linnæus, Haworth, Donovan vol 8, pl 280, Albin pl 5, fig 7, Wilkes Brit Butt, pl 11, Harris, Aurelian, pl 42, Lewin, pl 42.

Thecla Betulæ, Fabricius, Leach, Stephens, Curtis, Duncan pl 27, fig 1

Lycæna Betulæ, Ochsenheimer

Strymon Betulæ, Hubner (Verz bek Schm)

This species being the true Fabrician type of the genus Thecla, is placed at the head of the species of Hair-streak Butterflies; moreover it is the largest British species of the genus, the fore wings extending from $1\frac{1}{3}$ to rather more than $1\frac{1}{2}$ inches. Their upper surface is of a rich brown with a satiny gloss; the fore wings in both sexes are marked at the extremity of the discoidal cell with a short transverse black line generally succeeded in the males by an obscure orange cloud (which is, however, sometimes wanting), and which in the females is replaced by a large kidney-shaped orange patch; the cilia is whitish; the hind wings are marked at the anal angle and at the base of the tail (and of the succeeding lobe of the female) with orange; there is also a minute white spot near the anal angle. The under side of the wings is tawny yellow, with the edge brighter orange; at the extremity of the discoidal cell of the fore wings is a short transverse dark line edged with white, between which and the extremity of the wing is another broader dark orange wedge-shaped spot extending rather more than half-way across the wings, edged with a very slender dusky line which is margined with white on the outside. The hind wings are somewhat richer coloured, especially near the anal angle, with an abbreviated white line edged externally with a dusky line extending half-way across the middle of the wing; between this and the margin of the wing is another slender irregular white line, edged internally with a dusky line, the space between these two dusky lines being rich orange; the anal angle is marked with small black spots, and the cilia on each side of the tail is striped with brown.

The caterpillar is pale green with paler oblique lines along the sides, and straight ones down the back. It feeds on the birch, black-thorn, plum, &c. The chrysalis is brown, with darker marks. The perfect insect appears in the month of August. It is by no means a common species, although widely distributed. Coombe, Birch, Hornsey, and Darenth woods—Raydon wood, near Ipswich—Berkshire, Dorsetshire, Devonshire Dartmoor, and Norfolk—are given as the localities of this species.

SPECIES 2.—THECLA QUERCUS. THE PURPLE HAIR-STREAK BUTTERFLY.

Plate xxiv. fig. 5—10.

SYNONYMS.—*Papilio Quercus*, Linnæus, Lewin, P. p. pl. 43; Donovan, Brit. Ins. vol. 13, pl. 460; Wilkes, pl. 116; Harris Aurelian, pl. 16, fig. a—g; Albin, pl. 52, fig. a—c.
Thecla Quercus, Leach, Stephens, Curtis, Boisduval, Duncan, Brit. Butt. pl. 27, fig. 3—4; Westwood Introd. to Mod. Class. of Ins. 2. p. 357, fig. 100—8.
Bithys Quercus, Hubner.
Lycæna Quercus, Ochsenheimer.

This species varies in the expanse of its wings from $1\frac{1}{4}$ to $1\frac{1}{2}$ inches. On the upper surface the wings are of an obscure blackish-brown, more or less tinged in the males with a purple hue, which extends all over them, except along the hind margin; two varieties in the intensity of this purple tinge are represented in our plate. The female, on the contrary, (which is generally smaller than the male,) has the wings brownish-black, with a splendid glossy blue patch, which nearly covers the discoidal cell of the fore wings, and extends towards the anal angle of these wings (differing, however, in size in different specimens), the hind wings being immaculate. On the underside there is no difference in the markings between the two sexes, the general colour being of a pale ashy-brown, the apical portion being rather paler; beyond the middle of the fore wings is a transverse straight white slender bar, which does not extend to the anal angle, edged internally with brown, and between this and the margin is a row of darker spots, edged with grey, those near the anal angle being the largest and varied with fulvous; a more irregularly waved white line runs across the hind wings beyond the middle, also edged internally with brown, and between this and the margin of the wing are two rows of whitish-grey scallops terminating towards the anal angle in a fulvous eye, having a black pupil, and edged internally with black, the anal angle itself is also marked with fulvous and black. The tail of the hind wing is of a black colour, and not so long in this species as in T. Betulæ. The body is black above, ashy-grey beneath, the eyes are margined with white, and the club of the antennæ is reddish on the under side.

This species is found about the middle of July, flying over the tops of oaks. It occurs plentifully throughout England, but is rare in Scotland. The caterpillar also feeds on the oak, and is a thick oniscreform sluggish creature, clothed with short hairs, and with the upper surface of the body of a rosy hue with several rows of dark greenish lines or dots. In this state it offers several peculiarities worthy of notice, the head is small and entirely retractile beneath the first segment of the body, which is semicircular and margined with two dark scaly patches in the middle; this segment also bears a pair of spiracles, which are wanting in the two following segments, they exist however in the fourth, fifth, sixth, seventh, eighth, ninth, tenth, and eleventh segments. It is remarkable that the three terminal segments are soldered together without articulation on the upper side, although traces of articulation occur beneath. The feet are very minute. The anal prolegs appear placed on a distinct fleshy segment (Introd. to Mod. Class. Ins. ii. fig. 100, 9, 10.)

The caterpillar is found in the beginning of June. The chrysalis is of a shining rusty-brown colour, with three rows of brown spots on the back. A writer in Loudon's Magazine of Natural History (No. 32) states that the caterpillar of this species goes under ground to effect its transformations, but I believe its ordinary habit is to attach itself to the underside of the leaves of the oak.

SPECIES 3.—THECLA PRUNI. THE BLACK HAIR-STREAK BUTTERFLY.

Plate xxv. fig. 6—10.

SYNONYMES.—*Papilio Pruni*, Linnæus; Hubner.
Thecla Pruni, Stephens, Illustr. Haust. vol. 2, p. 66, note (not vol. I, p. 77, which is W. Album,) Curtis, Brit. Entomol. pl. 264; Duncan Brit. Butt. pl. 28. fig. 1.
Thecla Spini, Brit. Butt. p. 30.
Strymon Pruni, Hubner (Verz. bek. Schm.)
Lycæna Pruni, Ochsenheimer.

This species measures from $1\frac{1}{2}$ to nearly $1\frac{3}{4}$ inches in the expansion of its wings, which are of brownish-black colour, the anterior in the males having a small silky oval patch near the middle towards the costal margin: the posterior wings have two or three (and sometimes more) orange-coloured lunular spots near the hind margin towards the anal angle, where there is a small bluish dot occasionally, as in the upper figure of the twenty-fifth Plate. The orange lunules exist along the entire margin of the hind wings, and extend into the fore wings, but such is of rare occurrence, if indeed it be not the character of the female, as stated by Mr. Stephens. Beneath the ground colour is of a lighter brown, having an ochre-tint; the fore wings having a slender nearly straight bluish-white line extending across the wing beyond the middle, and reaching to the inner margin of the hind wings, where it assumes a more irregular appearance, somewhat resembling an obtuse W; beyond this line the fore wings are marked with several obscure fulvous patches, those nearest the anal angle being preceded by a small black and silvery dot or eyelet; these black spots, seven in number and edged internally with silver, are more conspicuous on the hind wings, and are succeeded by a broad fulvous bar extending to the anal angle, the outer edge of which is marked with semicircular black marks (followed by a silvery line), those nearest the anal angle being the largest; the anal angle itself is black with a silvery dot. The cilia is black at the base, and externally silvery with black spots: the tails are black, the antennæ are annulated with white, and the eyes are margined with the same colour.

This species has only been known as a native species during the last ten or twelve years. The earliest notice of its occurrence having been given by Mr. Stephens, as above referred to (and not in vol. ii. p. 69 of the Illustrations, as referred to by Mr. Stephens in vol. iv. p. 382). Shortly afterwards it was figured by Mr. Curtis, in his illustrations, who states that a number of specimens had been taken in Yorkshire by Mr. Seaman, in the preceding July (1828). Mr. Stephens states, however, that this locality is erroneous, and that the insect occurs in profusion in Monk's Wood, Herts. It is true we find a Thecla Pruni, in English works published previous to this period, but the fact was that English entomologists had mistaken the next species (Thecla W-Album) for the Linnæan Papilio Pruni, until the capture of the real species enabled them to correct their error. Mr. Bree informs us that this butterfly has been taken by his son in great abundance in Barnwell-wold, in July, 1838.

The caterpillar is green, with oblique yellowish lines at the sides, and darker marks down the back. The chrysalis is brown with lighter markings, and dark tubercles.

DESCRIPTION OF PLATE XXVI.

INSECTS.—Fig. 1. Thecla W-album, male (the white W-hair-streak Butterfly). 2. The female. 3. Showing the under side. 4. The Caterpillar. 5. The Chrysalis.

Fig. 6. Thecla Rubi, male (the green hair-streak B.). 7. The female. 8. Showing the under side. 9. The Caterpillar. 10. The Chrysalis.

PLANTS.—Fig. 11. Ulmus campestris (the common Elm).

„ Fig. 12. Rubus cæsius (the common Dewberry).

T. W-album and T. Rubi, are both from specimens in the British Museum; the Caterpillar and Chrysalis of the former are from Godart; of the latter, from Hubner. H. N. H.

SPECIES 4.—THECLA W-ALBUM. THE W-HAIR-STREAK BUTTERFLY.

Plate xxvi fig 1—5

SYNONYMS.—*Papilio W. Album*, Villers, Ent. vol 2, pl 4, fig 12
Thecla W. Album, Hubner, Godart, Stephens, Illust. vol 2, p 66, Curtis, Duncan, Brit. Butt. pl 28, fig 2
Papilio Pruni, Lewin, Pap pl 44, Haworth, Donovan, Brit. Ins. vol 13, pl 437
Thecla Pruni, Leach, Jermyn, Stephens' Illust. vol 1, p 77
Strymon W-Album, Hubner
Lycæna W. Album, Ochsenheimer

This species is closely allied to the preceding, but may at once be distinguished by the want of the orange marks on the upper side of the wings, and the more acute form of the W near the anal angle of the hind wings beneath, whence the name of the species. The expansion of the wings varies from a little less than 1¼ to rather more than 1½ inches. The upper surface of the wings is of a uniform dark brown or blackish, with a minute white, and sometimes a few rufous scales near the anal angle. The males have an oval glabrous spot near the middle of the fore wings towards the costa. On the under side, the ground colour of the wings is of a paler brown, and the fore wings are marked beyond the middle with a transverse white line (which is rather broader and more wavy in the female than in the male), and which does not extend to the anal angle; the hind wings are traversed by a slender white line beyond the middle, which is more slender and greatly angulated near the abdomen, forming the letter W. A row of slender black lunules (slightly edged internally with white) runs nearly parallel with the outer margin of the hind wings, succeeded by a fulvous band extending from the anal angle about half way towards the outer angle of this pair of wings, where it becomes gradually obliterated; externally this band is marked with black semicircular spots succeeded by a silvery line at the base of the cilia, (those nearest the anal angle being largest;) the anal angle itself is black with a silvery dot. The tails are black, tipped with white, those of the females being the longest; the antennæ are ringed with white, the tip reddish, the tarsi whitish ringed with brown.

The caterpillar is green, the posterior segments of the abdomen being spotted with dark red, and two rows of small dots down the middle of the back, which is dentated, and paler oblique lateral marks. Previous to undergoing its transformations, it assumes a brown colour. The elm and the black-thorn have been given as its food. The chrysalis is brown, with a white head.

Until about twelve years ago this was a scarce insect, and was confounded with the preceding.* In July 1827, however, Mr Stephens found it in myriads enlivening the hedges for miles in the vicinity of Ripley (but not to the north nor north-west of the village, although the bramble, upon the blossoms of which it chiefly delighted to settle, was in equal profusion there). Of their astonishing numbers then observed, an idea may be obtained when it is stated that he captured nearly 200 specimens in less than half an hour as they approached the bramble-bush near which he had stationed himself. This is the more remarkable, as he had never previously observed it in that neighbourhood, although he had frequently collected there; near Windsor, Cambridgeshire, near Ipswich and Bungay, Suffolk, and Southgate, Middlesex, have been given as additional localities. The Rev. W. T. Bree also informs us that his son has taken it sparingly in Barnwellwold and that it occasionally occurs near Allesley.

* "Papilio W album, Villars 2, 83, t 4, f 12, a P Pruni nullo modo differt." Haworth Lep Brit p 38

DESCRIPTION OF PLATE XXVII

INSECTS.—Fig 1 Thecla Spini, male 2 The female 3 Showing the under side 4 The Caterpillar 5 The Chrysalis

" Fig 6 Thecla Ilicis, female 7 Showing the under side 8 The Caterpillar 9 The Chrysalis

PLANTS.—Fig 10 Prunus Spinosa (Sloe)

Fig 11 Quercus Ilex

It is doubted by many, whether either of the above-named insects is British, but as they are said to have been captured upon one or two occasions, this work would not be complete without them T Spini is from a pair of German specimens in the British Museum T Ilicis from the figure of Hubner, and the larvæ of both are from the same source H N H

SPECIES 5.—THECLA SPINI THE PALE BROWN HAIR-STREAK BUTTERFLY

Plate xxvii fig 1—5

SYNONYMES.—*Hesperia Spini*, Fabricius, Hubner, Pap pl 75, fig 376, 377

Papilio Spini? Haworth, Ent Trans 1, p 336

Thecla Spini, Jermyn, Stephens, Wood Ind Entomol fig 53, Curtis?

Papilio Lynceus Esper 1 pl 39, fig 3

Strymon Spini, Hubner

Lycæna Spini, Ochsenheimer

In the third part of the Transactions of the old Entomological Society of which the late Mr Haworth was the main support, we find in a list of species of butterflies which had been exhibited from time to time before the Society, one called the pale brown hair-streak with the trivial name Papilio Spini attached, but accompanied by a mark of interrogation This specimen was stated by Mr Haworth to Mr Curtis to have been purchased in an old English collection Mr Stephens, however gave the Thecla Spini as an undoubted native species, on the further authority of a specimen which he states to have been taken in Norfolk by Mr Sparshall * He however gives a translation of the Fabrician character of T Spini instead of a description either of Mr Haworth's or Mr Sparshall's specimens, which is the more to be regretted as Mr Curtis states not only that Mr Sparshall received his specimen from some of his correspondents in town (thus rendering its real indigenousness questionable), but that Mr Haworth's specimen neither agreed with Hubner's figure of Spini nor with Mr Stephens (that is the Fabrician) description of the same species I regret that I have not been able to obtain access to either of these two reputed specimens for comparison, in consequence of the deaths of both Mr Haworth and Mr. Sparshall Mr. Haworth, however, allowed Mr Wood to make a drawing of his specimen, the under side of which is accordingly represented in the Index Entomologicus pl 2, fig 53, of the natural size, the fore wings measuring about an inch in expanse, the ground colour is pale brown, the fore pair having a transverse, straight, white streak beyond the middle of the wing extending from the costa, but not reaching the anal angle, and edged internally with a black streak A similar white streak edged within with black extends across the hind wings to the anal edge, where it forms a letter W more obtuse than in Th W-Album, but more clearly marked than in T Pruni The anal angle is brown, with a fulvous marginal streak, and there is a large fulvous patch with a black dot at the base of the tail, which is much longer than in any other English species The base of the cilia is black The true Spini closely resembles T Pruni the upper surface of the wings being of a brown colour, the hind pair having several red spots next the margin The tip of the tail is white Beneath, the

---

* Many specimens of the Thecla Spini are stated by Mr Hoy, in London's Magazine of Natural History (vol 2, p 88) to have been taken by Mr Seaman, of Ipswich The insects in question were Th Pruni

wings are ashy, with a white streak which is slightly angulated at the anal angle Towards the posterior margin are several fulvous lunules marked with black, and a large bluish spot at the anal angle, with a terminal black spot

The caterpillar is green, with dorsal lines formed of yellow spots and with a black head, when more mature it becomes reddish The chrysalis is pilose, brown above, and ashy beneath This description, which is translated from the Fabrician character, together with our figures taken from German specimens of T Spini, will enable any one to determine the species in case it should be captured in this country which is very probable

---

SPECIES 6—THECLA ILICIS THE EVERGREEN OAK HAIR-STREAK BUTTERFLY

Plate xxvii fig 6—9

SYNONYMES —*Papilio Ilicis*, Esper Schmetterl I t 39 Suppl 15 fig 1, b (fem) Borkhausen Eur Schm 1, p 138, 267, and 2 p 216, Hubner Pap t 75, fig 378, 379, Ochsenheimer, Schmett v I art I pt 2, p 103

Hesperia Rur Lynceus Fabricius, Ent Syst 3, p 279
Polyommate Lyncée, Godart Lep d France, 1, 186

The expansion of the wings of this species is rather greater than that of Th Spini The upper side of the wings in the male is of a blackish brown colour with a greenish tinge, whilst the female has a large orange patch on the disk of the anterior pair of wings beyond the middle In both sexes the anal angle is marked with a reddish spot The under side is brown grey, with a very slender somewhat waved white streak, edged internally with black, extending across all the wings and forming an obtuse W near the anal angle The hind wings are also ornamented with a row of orange crescents between the white streak and the margin of the wing, bordered with black, the anal angle itself being also orange, the tails and base of the ciliæ of the hind wing being black, edged internally with white

The caterpillar is green, varied with slender oblique lateral lines and dots, and a dark purple line down the side It feeds upon the evergreen oak

I have introduced this species on the authority of a specimen in the collection of Mr Meynell, stated to have been taken in Yorkshire From later information, however, given to me by Mr Stephens, there is perhaps reason to doubt the fact of the capture of the specimen in this country it having been purchased from a dealer in London who is reputed not to be sufficiently precise in the localities of his specimens of rare Lepidoptera, and who moreover deals in exotic and European as well as British insects It is proper to mention the circumstance, although there seems no reason why the species should not be a native of this country, the Quercus Ilex, or Common Evergreen Oak, which is a native of the south of Europe, having been extensively cultivated in Britain from a very remote period

---

SPECIES 7—THECLA RUBI THE GREEN HAIR-STREAK BUTTERFLY

Plate xxvi fig 6—10

SYNONYMES —*Papilio Rubi*, Linnæus, Lewin Pap pl 44, Haworth, Donovan, Brit Ins vol 13, pl 113 Wilkes, pl 118 Harris Aurelian, pl 26, figs a b, d g Albin, pl 5, fig 8

Thecla Rubi, Leach Stephens, Curtis, Duncan, Brit Butt pl 28, fig 3
Lycus Rubi, Hubner, (Verz bek Schmett)

This species, which from the nature of the food of its larva is more frequently observed flying nearer the ground than its congeners, varies in the expanse of its wings from one inch to an inch and a third, on the upper side the wings are of a uniform obscure brown colour, with a slight silky gloss, especially in the male, which has an oval opake spot near the middle of the wing towards the costa, the base of the wings has also a slight greenish tinge, on the under side the wings are of a uniform pea-green, except along the inner edge of the fore wings, which is hidden by the fore edge of the hind ones, and is of a brownish colour The hind wings are marked beyond the middle with a row of minute white dots, which vary in their size, being sometimes obsolete, and sometimes so large as to form a streak across the wings The tail is obsolete, its place being indicated by a slight projection or tooth, besides which there are several others, so that this pair of wings is denticulated The ciliæ are brown, dotted with black in the hind wings In addition to the variation in the size of the white dots on the hind wings beneath, Mr Stephens mentions a variety in which the fore wings on the under side have a row of white dots on the front margin, the female has also occasionally a pale whitish oval dot near the middle of the fore wings towards the costa

The caterpillar is pubescent, light green, with lateral rows of triangular yellow spots, and a white line above the legs The head is black It feeds on the bramble, broom, dyers'-weed, &c, and may be found at the middle of July This butterfly, which flies over white-thorn hedges, and especially bramble bushes and other low shrubs on which the caterpillar feeds, further differs from the rest of the genus in being double-brooded the first brood appearing in May or at the beginning of June, and the second at the beginning of August It appears to be distributed over the greater part of our island it has, however, only been observed in the southern counties of Scotland

GENUS XXII

CHRYSOPHANUS, HUBNER, (POLYOMMATUS, Boisduval, LYCÆNA Stephens)

This genus, restricted to the butterflies which are termed Coppers by collectors, is distinguished from the other species of the family not only by the brilliant colour of the upper surface of the wings, but by having the antennæ long and terminated by an abrupt fusiform club, which is not spoon-shaped, the hind wings more denticulated than in the Blues, but destitute of the tails the pulvilli of the feet are also larger than in the last-named genus, whilst the naked eyes separate the species from those of Thecla In the males of most of the species the hind wings have the anal angle produced, and in the females the hind margin of the wing jointing the angle is subemarginate The palpi are nearly straight, with the last joint naked, rather long and subulated, and the head is narrower than the thorax The postcostal vein of the fore wings emits three branches extending to the costa, the third of which, arising near to, or rather beyond the union of the postcostal with the ordinary transverse vein is forked, as in Thecla Quercus, and Betulæ Boisduval's figure of the veins of this genus, Hist Nat Lep 1, pl 6—C, fig 7, is (as I am sorry to say many of his other figures of the veins are) inaccurate The ground colour of the wings above is fiery orange, at least in one sex, and the females have the upper side of the wings always marked with black spots

The caterpillars resemble rather elongated woodlice, and appear somewhat hairy when seen through a lens They feed on low plants Dr Horsfield indeed considers that the chief difference between his genera Polyommatus and Lycæna depends on the variation of the metamorphosis, the larva in the former being regularly rounded or cylindrico-gibbose, in the latter more oblong and unpressed at the sides He, in fact, states that in the antennæ and palpi of these two genera no tangible difference can be pointed out the distinction derived from the wings is however more decisive, although the distinction, though not easily described in words is readily seized by an experienced eye The Blues are altogether without tails, and their character is well preserved in the oriental tropical regions, but it is remarkable that in that part of the world no true Coppers have been discovered, which, in Europe, chiefly constitute the present genus

The relationship of this genus with that of the Blues is indeed very close, and we accordingly find considerable diversity of opinion as to the employment of the generic names of the two groups Latreille in all his works employed the name Polyommatus for the whole of the species of the present family, giving at the first our Thecla as the primary division, whilst in his late works he gives one of the Blues as an example of the genus Fabricius separated some of the Hair-streaks under the name of Thecla, retaining the Blues and Coppers, together with a great number of foreign species, under the name of Lycæna, which name Ochsenheimer employed, arranging, however, the entire family under this generic name, and forming the Blues, Coppers and Hair-streaks into three groups or families Mr Curtis, in his British Entomology, gave the Blues and Coppers under the genus Lycæna, but in his Guide he has adopted the nomenclature of Mr Stephens, namely Thecla for the Hair-streaks, Lycæna for the Coppers, and Polyommatus for the Blues

Dr Horsfield, in his beautiful work on the Lepidoptera of Java (which, it is to be hoped, will still be completed, at least as far as regards the diurnal Lepidoptera), placed in the genus Polyommatus only two Javanese species allied to P Alsus and P Argiolus, but greatly enlarged the limits of Lycæna, making it comprise many Blues, whilst Boisduval adopts the name Lycæna for exotic insects, such as P Bœticus, Linn, calling our Blues by the name of Argus, taken up from Scopoli and Geoffroy, and our Coppers Polyommatus

Such is one of the many instances of confusion to be met with in the works of modern entomologists, owing to the want of some fixed principle regulating the adoption of old generic names when the genera are required to be cut up into minor groups The names both of Lycæna and Polyommatus were as we have seen, intended to indicate groups of greater extent than our present genera If, therefore, we form our Coppers into one group, and our Blues into another, the generic names Lycæna and Polyommatus, as intended by their original proposers, are not applicable thereto, unless indeed we can ascertain that they were regarded by the proposers of such names as their types and we have seen that such is not the case with either name, the Hair-streaks being placed at first at the head of Polyommatus by Latreille, whilst we find the Coppers at the end of the genus Lycæna of Fabricius The Purple Emperor, again, is placed at the head of the Argus group by Geoffroy, which also includes the Blues and the Coppers In such cases my opinion is that (in order to avoid such distracting confusion) wherever a species or division of a genus is separated from an old genus, a new name ought to be given to it unless such species or division be the true type of the old genus, when, of course, it will retain the old generic name Instances might be pointed out in which entomologists are agreed as to the nomenclature of a group, although, from the non-adoption of some such principle as this an old generic name has been abstracted from the true type of a genus and conferred upon an aberrant species, but in the case of the Copper and Blue Butterflies no such uniformity of opinion prevails, each writer having acted without any principle In the present instance, therefore, I feel no hesitation in rejecting the nomenclature of recent Lepidopterologists, being convinced that a revision of the entire family Lycænidæ will necessitate the establishment of a much greater number of named groups, when the name of Lycæna will have to be restored to the true type of the genus I have, therefore, adopted Hubner's name Chrysophanus for the present group, which is quite expressive of their splendid appearance, being derived from the Greek words χρυσός, gold, and φαίνω, to appear

(*To be completed in about Sixteen Numbers.*)

No. XI.] [PRICE 2s. 6d.

BRITISH INSECTS AND THEIR TRANSFORMATIONS.

BRITISH BUTTERFLIES

AND

Their Transformations.

ARRANGED AND ILLUSTRATED IN A SERIES OF PLATES

BY H. N. HUMPHREYS, ESQ.

WITH CHARACTERS AND DESCRIPTIONS

BY J. O. WESTWOOD, ESQ., F.L.S.,

SEC. OF THE ENTOMOLOGICAL SOCIETY, ETC. ETC.

LONDON:
WILLIAM SMITH, 113, FLEET STREET.

MDCCCXLI.

BRADBURY AND EVANS, PRINTERS, WHITEFRIARS.

WORKS PUBLISHED BY ...

SMITH'S STANDARD LIBRARY.

In medium 8vo, uniform with Byron's Works, &c

Works already Published

ROKEBY BY SIR WALTER SCOTT 1s 2d

THE LADY OF THE LAKE 1s
THE LAY OF THE LAST MINSTREL 1s
MARMION 1s 2d
THE VISION OF DON RODERICK, BALLADS AND LYRICAL PIECES By SIR WALTER SCOTT 1s
THE BOROUGH By the REV G CRABBE 1s 4d
THE MUTINY OF THE BOUNTY 1s 4d
THE POETICAL WORKS OF H KIRKE WHITE 1s
THE POETICAL WORKS OF ROBERT BURNS 2s 6d
PAUL AND VIRGINIA, THE INDIAN COTTAGE, and ELIZABETH 1s 6d
MEMOIRS OF THE LIFE OF COLONEL HUTCHINSON, GOVERNOR OF NOTTINGHAM CASTLE during the Civil War By his Widow, Mrs LUCY HUTCHINSON 2s 6d
THOMSON'S SEASONS, AND CASTLE OF INDOLENCE 1s
LOCKE ON THE REASONABLENESS OF CHRISTIANITY 1s
GOLDSMITH'S POEMS AND PLAYS 1s 6d
———— VICAR OF WAKEFIELD 1s
———— CITIZEN OF THE WORLD 2s 6d
———— ESSAYS, &c 2s

, These four Numbers form the Miscellaneous Works of Oliver Goldsmith, and may be had bound together in One Volume, cloth lettered, price 8s

KNICKERBOCKER'S HISTORY OF NEW YORK 2s 3d
NATURE AND ART. By MRS INCHBALD 10d
A SIMPLE STORY By MRS INCHBALD 2s
ANSON'S VOYAGE ROUND THE WORLD 2s 6d
THE LIFE OF BENVENUTO CELLINI 3s
SCHILLER'S TRAGEDIES, THE PICCOLOMINI, and THE DEATH OF WALLENSTEIN 1s 8d
THE POETICAL WORKS OF GRAY & COLLINS 10d
HOME By MISS SEDGWICK 9d
THE LINWOODS By MISS SEDGWICK 2s 8d
THE LIFE AND OPINIONS OF TRISTRAM SHANDY By LAURENCE STERNE 3s
INCIDENTS OF TRAVEL IN EGYPT, ARABIA PETRÆA, AND THE HOLY LAND By J L STEPHENS, Esq 2s 6d
INCIDENTS OF TRAVEL IN THE RUSSIAN AND TURKISH EMPIRES By J L STEPHENS, Esq 2s 6d.
BEATTIE'S POETICAL WORKS, and BLAIR'S GRAVE 1s.
THE LIFE AND ADVENTURES OF PETER WILKINS, A CORNISH MAN 2s 4d
UNDINE A MINIATURE ROMANCE Translated from the German, by the REV THOMAS TRACY 9d
THE ILIAD OF HOMER Translated by ALEX POPE 3s
ROBIN HOOD, a Collection of all the ancient Poems, Songs, and Ballads now extant, relative to that celebrated English Outlaw, to which are prefixed, Historical Anecdotes of his Life Carefully revised, from RITSON 2s 6d
THE LIVES OF DONNE, WOTTON, HOOKER, HERBERT, AND SANDERSON Written by IZAAK WALTON 2s 6d
THE LIFE OF PETRARCH By MRS DOBSON 3s
MILTON'S PARADISE LOST 1s 10d
———— PARADISE REGAINED, and MISCELLANEOUS POEMS 2s
RASSELAS By DR JOHNSON 9d
ESSAYS ON TASTE By the Rev ARCHIBALD ALISON, LL B 2s 6d
THE POETICAL WORKS OF JOHN KEATS 2s
THOMSON'S POEMS AND PLAYS, COMPLETE 5s
PICCIOLA A Tale from the French 1s 4d
THE POETICAL WORKS OF DR YOUNG, COMPLETE 5s

FOUR VOLUMES ARE NOW ARRANGED ON THE FOLLOWING PLAN —

ONE VOLUME OF "POETRY,"

CONTAINING

SCOTT'S LAY OF THE LAST MINSTREL
SCOTT'S LADY OF THE LAKE
SCOTT'S MARMION
CRABBE'S BOROUGH
THOMSON'S POETICAL WORKS
KIRKE WHITE'S POETICAL WORKS
BURNS'S POETICAL WORKS

A SECOND VOLUME OF "POETRY,"

CONTAINING

MILTON'S POETICAL WORKS
BEATTIE'S POETICAL WORKS
BLAIR'S POETICAL WORKS
GRAY'S POETICAL WORKS
COLLINS'S POETICAL WORKS
KEATS'S POETICAL WORKS
GOLDSMITH'S POETICAL WORKS

ONE VOLUME OF "FICTION,"

CONTAINING

NATURE AND ART By MRS INCHBALD
HOME By MISS SEDGWICK
KNICKERBOCKER'S HISTORY OF NEW YORK By WASHINGTON IRVING
PAUL AND VIRGINIA By ST PIERRE
THE INDIAN COTTAGE By ST PIERRE
ELIZABETH By MADAME COTTIN
THE VICAR OF WAKEFIELD By OLIVER GOLDSMITH
TRISTRAM SHANDY By LAURENCE STERNE

ONE VOLUME OF "VOYAGES AND TRAVELS,"

CONTAINING

ANSON'S VOYAGE ROUND THE WORLD
BLIGH'S MUTINY OF THE BOUNTY
STEPHENS'S TRAVELS IN EGYPT, ARABIA PETRÆA, AND THE HOLY LAND
STEPHENS'S TRAVELS IN GREECE, TURKEY, RUSSIA, AND POLAND

These Volumes may be had separately, very neatly bound in cloth, with the contents lettered on the back, price 10s 6d each

Those parties who have taken in the different works as they were published, and who wish to bind them according to the above arrangement, may be supplied with the title-pages gratis, through their booksellers, and with the cloth covers at 1s each

DESCRIPTION OF PLATE XXVIII

INSECTS.—Fig. 1. Chrysophanus Chryseis (the purple-edged Copper Butterfly). 2. The Female. 3. Showing the under side.
" Fig. 4. Chrysophanus Phlæas (the common Copper Butterfly). 5. Showing the under side. 6. 11. Caterpillar. 7. The Chrysalis.

PLANTS.—Fig. 9 and 10. Rumex acetosella (the Sorrel).

The most brilliant though perhaps not the most beautiful of this genus is here figured with the most common and least striking of its family, the well known little Meadow Copper. I have not given the female, as, unlike the rest of the genus, it does not differ from the male, but instead have figured a beautiful variety described by Hubner, in which all the dark brown and black marks are extracted, leaving creamy white in their place, whilst the orange marks remain perfect. This beautiful variety of the common Copper has been taken in England, and it might be worth the experiment of collectors to rear a number of broods for the chance of obtaining a specimen. The Caterpillar and Chrysalis are from Godart.—H. N. H.

SPECIES 1.—CHRYSOPHANUS PHLÆAS. THE COMMON COPPER BUTTERFLY.

Plate xxviii. fig. 4—8.

SYNONYMES.—*Papilio Phlæas*, Linnæus, Haworth, Lewin, pl. 41, Donovan, vol. 13. pl. 466. Harris, Aurelian, pl. 34.
Lycæna Phlæas, Fabricius, Ochsenheimer, Leach, Stephens, Curtis, Duncan Brit. Butterflies pl. 30, fig. 3. Wood, Ind. Entom. t. 2. fig. 56.
Chrysophanus Phlæas, Hubner (Verz. bek. Schmett.)
Polyommatus Phlæas, Boisduval.

This very pretty, and, at the same, abundant species varies in the expanse of its wings from 1 to 1½ inches. The fore wings on the upper side are of a shining fiery copper colour, ornamented on the disk with from eight to ten black spots of unequal size and dissimilar shape, of which the three or four nearest the extremity of the wing are placed transversely, and more or less confluent, and are preceded by a detached spot which is larger than the rest. The front margin of the wing is narrowly, and the hind margin broadly edged with brown; the hind wings above are dark brown powdered at the base with copper, with several nearly obsolete black marks on the disc; near the hind margin is a bar of copper of variable breadth, edged above and beneath with black spots so as to cause the bar to appear as if formed of five confluent patches. The upper row of these black spots is often preceded by blue irrorations. Beneath, the fore wings are of a fulvous colour without any gloss, the black spots being more distinct, and slightly edged with buff, one near the base of the wings, and a minute one on the costa near the tip being on this side quite distinct; the hind margin is drab coloured, with several dark-coloured crescents next the anal angle; the under wings on this side are also drab-coloured, with numerous minute, nearly obsolete brown marks placed transversely, and with an obscure narrow orange band parallel with the hind margin. The tails in this species are longer than in the others of the genus. The body above is black, with tawny hairs about the head and thorax, and drab-coloured beneath. There is no difference in the colour and markings of the two sexes.

The perfect insect is distributed throughout the country, and appears at the beginning of April, June and August; thus there are several broods in the course of the year. The caterpillar is green, with a pale dorsal and lateral line, and feeds on the sorrel. I have received specimens from North America.* It also occurs throughout Europe and in Asia.

---

* It is proper to observe, that my American specimen differs in the decided black spotting of the under side of the hind wings, the bright red streak near their hind margin, and in wanting the minute spot on the costa of the fore wings; but these characters can scarcely be held to constitute a distinct species.

This species, in the similarity of the markings in the two sexes, the greater length of the tails of the hind wings and in the circumstance of there being several broods in the course of the summer, is aberrant from the general characteristics of the genus. This is one of our most elegant and active butterflies, frequenting pasture-lands, commons, heaths, &c., and keeping up a continual warfare with its fellows. Its beautiful appearance in the glowing sunshine, especially when contrasted with the colours of the flowers on which it delights to settle, renders it conspicuous to every passenger.

There is considerable variation in the intensity and extent of the copper markings of this species. In addition to these ordinary variations, three other more striking varieties have been found: one, in which the copper band of the hind wings is quite obliterated; another, in which the copper colour on both surfaces of the wings is replaced by milk-white (of which variety there is a specimen in the collection of the Entomological Society of London from the cabinet of Mr. Kirby); and a third (figured in our plate 28, fig. 8), in which, on the contrary, it is the black portion of the wings which is replaced by milk-white.

SPECIES 2.—CHRYSOPHANUS CHRYSEIS. THE PURPLE-EDGED COPPER BUTTERFLY.

Plate xxviii. fig. 1—3.

SYNONYMS.—*Hesperia Chryseis*, Fabricius.
Papilio Chryseis, Haworth; Sowerby, British Miscell. 1, pl. 13.
Lycæna Chryseis, Ochsenheimer; Stephens; Duncan, Brit. Butt. pl. 30, fig. 1; Wood, Ind. Entom. t. 2, f. 37.
Chrysophanus Chryseis, Hübner (Verz. bek. Schmett.).
Papilio Hippothoe, var., Esper, pl. 62, fig. 1.
Pap. Eurydice, Borkhausen.

The expansion of the wings of this handsome species is about two inches and a half. The upper side of the wings in the male is of a shining copper colour, of a redder tint than in C. Phlæas. At the extremity of the discoidal cell in all the wings is a slender black bar; the hind margin is brown, within which, as well as along the front edge of the fore wings, is suffused a rich purple tint, which extends also along the anal edge of the hind wings, which in that part have the ground colour of a brown hue, which extends in a broad margin along the extremity of these wings, having an ill-defined series of copper spots near the anal angle. The female, on the contrary, has the disc of the fore wings, above, of a dull copper colour without any gloss, and with the edges brown; there is also a dot in the middle, and a bar at the extremity of the discoidal cell, a curved bar of six brown spots beyond the middle of the fore wings, and a more indistinct row of smaller spots nearer the margin; the disc of the hind wings is entirely brown, with a narrow bar of dull orange near the anal angle, spotted with brown both above and below. Beneath, both sexes are alike, except that the fore wings of the female have the disc more suffused with orange. The ground colour of all the wings on this surface is ashy drab, with the margins more grey, and the base more slate-coloured or bluish; the fore wings have about seventeen black spots of variable size, ocellated with white; the three anterior ones placed longitudinally, the third being transverse, succeeded by seven in a curved series, the remainder more indistinct, and running parallel with the extremity of the wing; each of the hind wings is marked with about thirty similar ocellated spots, those at the base of the wings being scattered about, whilst the others are arranged in transverse curved bands, those upon the margin of the wing being almost obsolete; between these and the preceding row of spots, the wings are dashed with orange near the anal angle. There is, however, considerable diversity in the ground colour of the wings beneath, as well as in the number of spots; the specimen figured, for example, in the upper part of our plate 28 has fewer than the ordinary number of ocellated spots.

This is one of the rarest British species, indeed, by some collectors its claim to be considered an indigenous insect is considered as doubtful Mr. Stephens says that Dr Leach received fine and recent specimens from the vicinity of Epping for several successive seasons I believe, however, they were obtained from a dealer, who persisted in keeping the precise locality secret This, of course, he would have done whether the specimens were native or obtained from abroad as it would have diminished their value, if British had other collectors been made acquainted with the spot It is also said to have been taken in Ashdown Forest in Sussex It appears at the end of the summer, frequenting marshy places

DESCRIPTION OF PLATE XXIX

INSECTS —Fig 1 Lycæna dispar, male (the large copper Butterfly) 2 The female 3 Showing the under side 4 A common variety of the female 5 The Caterpillar 6 The Chrysalis

PLANTS —Fig 7 Iris Pseud-acorus (the yellow water Iris) 8 Rumex palustris (the Marsh Dock)

I have devoted an entire plate to this, perhaps the handsomest, of British butterflies, which is moreover, interesting as an example of the results of research in natural history It was actually unknown to naturalists till very few years ago although the fens in Huntingdonshire and parts of Cambridgeshire absolutely swarm with it in the month of July It was supposed to be an exclusively British insect, and Hubner has figured it from English specimens But I have a specimen which I took between Rome and Naples, in the Pontine Marshes, which appears perfectly identical in every respect with English specimens The caterpillar I have been enabled to publish for the first from a drawing by Mr Stephens, which he has obligingly furnished me with for the purpose —H N H

SPECIES 3 —CHRYSOPHANUS DISPAR THE LARGE COPPER BUTTERFLY

Plate XXIX fig 1—6

SYNONYMES —*Papilio dispar*, Haworth Kirby and Spence Introd to Ent I pl 3 f 1 male

Lycæna dispar, Curtis, Brit Ent 1 pl 12, Duncan, Brit Butt pl 20, fig 1—2 Wood Ind Ent t 3 f 59 male and fem Swainson Zool Illust n ser, pl 132

Polyommatus dispar, Boisduval Icon Histor Lepid I iii pl 10 fig 1—3

Papilio Hippothoe, Lewin, Ins pl 40 Donovan Ins pl 217

Papilio Hippothoe, var Esper Pap pl 11, f 1—2

This splendid species varies in the expanse of its wings from 1½ to rather more than 2 inches The upper surface of the wings in the male is of a brilliant fiery copper colour, similar in its tone to that of L Phlæas The fore wings are marked with a small black spot in the middle, and a transverse one at the extremity of the discoidal cell, between the latter and the outer margin of the wing are to be observed traces of the spots of the under side, the front margin and extremity of the fore wings are narrowly edged with black, which is broadest at the apex of the wing The hind wings have also a slender transverse mark at the extremity of the discoidal cell, between which and the hind margin are also traces of the rows of spots of the under side the hinder and anal margin are also black, on the margin are five black spots, the anal one being doubled The fore wings in the female on the upper side are of a darker copper colour, the base and fore margin being irrorated with brown, on the disc are eight black spots, one within and another at the extremity of the discoidal cell, the others forming a transverse bar beyond the middle of the wing the inner one being doubled The rudiment of another spot also appears near the base of the wing The black outer margin is broader than in the male The disc of the hind wings in this sex above is brown-black, more or less irrorated with copper, the veins being copper-coloured, running into a bar of copper near the hinder extremity of these wings the edge itself being brown with six triangular black-brown spots extending into the copper bar and giving it a lobed appearance

On the under side both sexes are alike the disc of the fore wings being pale fulvous with the edges ashy, with ten very distinct black ocellated spots, each with a slender pale iris three of these spots are placed longitudinally, the others forming a waved band across the wing, the two inner ones being small, more or less confluent this is succeeded by a row of obscure unequal sized dark spots The hind wings beneath are of a pale silvery blue, which becomes greyer as it recedes from the body, with a slender oblique bar at the extremity of the discoidal cell, and about twenty-five black spots, various in size, those towards the base of the wings being placed irregularly and ocellated with white, as well as the transverse irregular row formed of nine spots beyond the middle of the wing This is succeeded by a row of dark spots, followed by a submarginal fulvous bar, between which and the grey extremity of the wing is a very indistinct row of similar spots The caterpillar is described by Mr Stephens as "somewhat hairy bright green with innumerable white dots, it feeds upon a kind of dock The chrysalis is at first green, then pale-ash coloured, with a dark dorsal line and two abbreviated white ones on each side, and lastly sometimes deep brown " The fen districts of Cambridge and Huntingdonshire are the localities for this beautiful species, which appears not to be known as a native of any other part of Europe Benacre, Suffolk, and Bardolph fen, in Norfolk, have also produced it It is also said to have been taken by the botanist Hudson in Wales, but Mr Stephens thinks it probable that this locality belonged to P Hippothoe Donovan states that the specimens from which his figures are drawn were from Scotland, but Mr Haworth says, "Nunquam in Scotia ut amicus meus E Donovan ex informatione erronea dixerit "

Within the last twenty years the insect has become common in collections, owing to the immense numbers taken by collectors in the former localities, which, however, as I understand has almost extirpated the species

Boisduval, who at first gave it as Hippothoe, has since figured it under the name of Polyommatus dispar, observing, "ce joli Polyommatus n'est très probablement qu'une variété locale d'Hippothoe, remarquable par sa taille Il est au moins un tiers plus grand qu'Hippothoe du Continent, ses ailes sont d'un fauve plus vif, et elles ont souvent un reflet un peu purpurin "

The end of July and the beginning of August is the period of the appearance of this insect in the perfect state, it is found flying amongst the reeds growing in the fens, and is very active In Loudon's Magazine of Natural History, (No 37,) Mr Dale has noticed a variation in the form of the wings of this species

DESCRIPTION OF PLATE XXX

INSECTS —Fig 1 Lycæna Virgaureæ (the scarce copper Butterfly), male, 2 The female 3 Showing the under side 4 The Caterpillar 5 The Chrysalis

Fig 6 Lycæna Hippothoe (the dark under winged or tawny copper Butterfly) 7 The female 8 Showing the under side

PLANTS —Figs 9 and 10 Solidago Virgaurea (the Golden rod)

Hippothoe and Virgaureæ are sufficiently distinct on the upper surface, the former being of a pure tawny hue, whilst the latter presents an intense and brilliant copper colour, somewhat more orange and fiery in its tone than Dispar The under sides are very distinct, that of Hippothoe however, closely resembles that of Dispar The Caterpillar and Chrysalis are from Godart —H N H

SPECIES 4 —CHRYSOPHANUS HIPPOTHOE THE DARK UNDER-WINGED COPPER BUTTERFLY

Plate xxx fig 6—8

SYNONYMES —*Papilio Hippothoe*, Linnæus, Rosel, Ins Bel cl 2, t 37, f 6—7, male Haworth Lnt Trans p 333 (nec Lewin and Donovan)

Lycæna Hippothoe, Stephens, Curtis, Duncan, Brit Butt pl 30, fig 2 Wood, Ind Entomol pl 2, fig 58, male, pl 3, fig 58 female

Polyommatus Hippothoe, Boisduval

Chyrysophanus Hippothoe, Hubner (Verz bek Schmett)

This species, which has been regarded by many Entomologists as specifically identical with the preceding, differs from it in its constantly smaller size, the fore wings never expanding more than an inch and a half It is therefore, described by Linnæus as of the size of Virgaureæ. The wings of the male on the upper side are of a pure tawny or fulvous colour, with the outer edges alone black, and in the hind wings marked within with small black spots, of which the fourth is placed nearer the base of the wing. In most specimens there is also a transverse line at the extremity of the discoidal cell. The female has the upper surface of the fore wings dull copper with spots, arranged as in Ch. dispar, but smaller. The hind wings have the entire upper surface dusky without orange veins, but marked with darker spots, the margin itself being black and internally crenated. On the under side both sexes resemble each other, the disc of the wings being luteous-ash coloured, with fewer and smaller spots but similarly ocellated, as in Ch. dispar, three of larger size being placed along the discoidal cell longitudinally, succeeded by an irregular row placed transversely, and several very minute ones parallel with the outer margin; the hind wings beneath are ash-coloured, with about seventeen ocellated spots, and a fulvous band on the hinder margin anteriorly spotted with black.

Although regarded by some authors as specifically identical with Ch. dispar, the present species differs in its smaller size, more tawny hue of the upper side of the wings of the male (generally destitute of the small transverse bar or streak at the extremity of the discoidal cell of the fore wings), fewer and smaller spots on the under side of the wings, and more uniform hue of the hind wings of the female, which seem to warrant a specific distinction between the two insects; moreover, as Mr. Stephens observes, amongst several hundreds of Dispar which have been taken at Whittlesea Mere, not one specimen occurred agreeing with the true Hippothoe. Nothing is known with certainty as to the true locality of this species, of which several specimens were preserved in old English collections. It is presumed, however, that one of these was taken in some part of Kent, having having been obtained from an old collection made in that county, known to collectors under the name of the Kentish Cabinet.

SPECIES 5.—CHRYSOPHANUS VIRGAUREÆ. THE SCARCE COPPER BUTTERFLY.

Plate xxx, fig. 1—5.

SYNONYMES.—*Papilio Virgaureæ*, Linnæus; Haworth; Donovan, vol. 5, pl. 173, male; Lewin, pl. 41, f. 1—2. *Lycæna Virgaureæ*, Fabricius; Ochsenheimer; Stephens, Brit. Ent. 1, pl. 9, fig. 1—2, male, fig. 3, female; Duncan, Brit. Butt. pl. 29, fig. 3; Wood, Ind. Ent. t. 3, fig. 60, male and female. *Chrysophanus Virgaureæ*, Hubner (Verz. bek. Schmett.).

This very distinct species is about the size of Ch. Hippothoe, the fore wings expanding about an inch and a half; the upper surface of all the wings is of a very rich yellow copper colour, without any discoidal spots or any clouds indicating the situation of the spots on the under side of the wings; the margin of all the wings is black and more or less narrow, the hind wings having, moreover, a few black spots near the posterior edge, and confluent with the dark margin, except the two next the anal angle, which are close together and smaller than the rest. The female is more obscure in the colours of the upper side, with a spot in the middle and a larger one at the extremity of the discoidal cell; beyond this is an irregular row of black spots, which is succeeded by a submarginal row of six large somewhat confluent black spots. The hind wings are more variegated in their appearance than in any other species, being of an obscure fulvous colour at the base, a large discoidal patch and several dashes of brown between the veins of the wings (including a transverse curved row of seven larger spots nearly square, beyond the middle of the wing), and a marginal row of smaller ones of a dusky brown colour. Beneath the sexes are nearly similar, being of a dull fulvous stone colour, the base and extremity being irrorated with greenish. In the discoidal cell are two small black spots, and a transverse one at its extremity; beyond this are six or eight small black spots placed irregularly but in pairs, and beyond this the margin is clouded with a row of dusky spots. The posterior wings are dusky at the base, and they are marked with about twelve small black spots, those towards the base of the wing being placed irregularly, but those near the middle placed in a series across the wing, each with a small patch of white below it, so as unitedly to form an interrupted white bar. Near the anal angle are a few orange spots, and the angle itself is rather acute, and has an emargination adjoining it.

The caterpillar is pubescent and of a dull green colour, with a pale yellow line on the back and pale green streaks on the sides. It feeds on the golden rod, sharp dock (Rumex acutus), &c. The chrysalis is brownish yellow with dusky red wing-covers.

No specimens of this species having occurred for a great number of years, the claim of this insect to be regarded as indigenous has begun to be questioned. It has at all times been very rare. According to Lewin, two specimens of it were taken by himself in marshes, and Donovan states that one was once taken in Cambridge. The marshes in the Isle of Ely and Huntingdonshire are also stated as localities of this butterfly, which appears in the perfect state at the end of August. I possess a specimen given to me by the late Mr. Haworth as an undoubted native specimen.

POLYOMMATUS,* Latreille (or [illegible])

(*Polyommatus*, Stephens, Curtis, Horsfield, nec Boisduval; *Argus*, Geoffroy, Scopoli, Boisduval; *Lycæna*, [illegible] Ochsenh., Fabricius.)

Referring to the observations made under the genus Chrysophanus (pages 91 and 92), as to the close relationship existing between that genus and the present (which comprises the Blues of collectors), it will be sufficient in this place to observe that this genus is distinguished by having the upper surface of the wings generally of a blue colour, especially in the male, but occasionally brown in the females, with a row of fulvous spots near the outer margin; the under surface generally greyish with numerous ocelli, with black pupils surrounded by white irides.

The antennæ are filiform, and terminated by an abruptly-formed elongated compressed club terminating in a lateral point. The palpi are longer than the head, with the terminal joint naked and sharp. The fore legs have been described as alike in both sexes, but such is not the case (see *ante*, p. 81, note, and my Introd. to Mod. Class. of Ins., vol. ii, p. 358, fig. 100, 12, 13). The tarsi are furnished with minute simple ungues extending beyond the minute pulvilli. The wings are entire and without tails, the posterior being scarcely denticulated at the anal angle. The larvæ are onisciform, with the head and feet very small and scarcely perceptible, the body laciniate, the back elevated and generally beautifully coloured. The pupa is rather long, naked, and of a whitish colour, with some dusky spots on the back and sides. The species generally undergo their transformations on the stem of a plant, but occasionally beneath the surface of the earth.

The genus extends all over Europe. Species are also found in the north of Africa, the Cape of Good Hope, Madagascar, the Isles of France and Bourbon, the East Indies, and North America. Boisduval also mentions a species from New Ireland. I possess several species from New South Wales, and Captain Ross brought one from the Arctic regions. Mr. Swainson, however, informs us that they are almost unknown in South America.

* Πολυς, many, and ὄμμα, an eye, in allusion to the numerous ocelli on the under side of the wings of the genus.

Caterpillars of such species as have been observed feed upon leguminose herbs, such as Trifolium, Lotus Onobrychis, Medicago, &c

From the generic synonymes given above, it will be seen that the French, German and English schools of Entomology are at variance as to the name to be given to these insects It in this instance I have followed the authors of our own country, it is, first, because Latreille himself the founder of the genus has in his later works given one of the Blues as its type, secondly, because the name is a very expressive one, and thirdly, because the objections to the use of either of the names of Argus or Lycæna for the present group are as strong as those against the employment of that of Polyommatus

The number of the species in the genus being considerable, Ochsenheimer divided it artificially into two sections, according to the presence or want of a row of fulvous spots within the hind margin of the posterior wings beneath Subsequently Mr Stephens observed that "P Argiolus differs from its indigenous congeners by the form and texture of its wings, that P Alsus, Agestis and Artaxerxes are characterised by a uniformity of colouring in both sexes, while the remaining species are distinguished in general by the males being blue above, and the females brown, excepting Po Arion and Alcon, in which the latter sex is known by a predominance of brown above and by having the disk considerably spotted with dusky or black, and that the five first species (Argiolus, Alsus, Acis, Arion, and Alcon) are destitute of a marginal fascia beneath, which is, however, rudimentary in the two last-mentioned insects Again, some few of the species have the eyes pubescent while others have them naked' (Illustr B Ent Haust 1 p 85)

Dr Horsfield, in the "Lepidoptera Javanica," divided the genus Polyommatus into two subgenera, the first named Pithecops from the peculiar aspect of the chrysalis distinguished by a very distinct habit and aspect, 'owing to the great length and lateral expansion of the wings, to their comparative *narrowness* and to their (especially the posterior pair) being regularly elliptical and rounded in the anal region This subgenus is represented in the European Fauna by P Alsus, and several others described by Ochsenheimer having the character 'alæ integerrimæ' The subgenus Polyommatus, properly so called is characterised by Dr Horsfield by having the margins of the hinder wings with the anal extremity angular, and produced to a short, rounded point Mr Stephens, in his subsequently published Catalogue, adopts these two subgenera as sections giving Argiolus and Alsus, as well as Acis, as belonging to Pithecops, and in his manuscripts, which he has been so kind as to allow me to examine, he confines Pithecops to Argiolus giving Alsus, Acis, Arion and Alcon under the sectional name of Nomiades, and the remainder under that of Agriades, from Hubner As, however, Dr Horsfield gives Alsus expressly as the European type of Pithecops, which he characterises by the comparative narrowness of the wings, and as Argiolus has broader wings than any other European species, we must restrict Pithecops to P Alsus, which species, indeed, possesses a peculiarity in the arrangement of the veins of the fore wings which has not hitherto been noticed, and which I have found in no other Lepidopterous insect, thus confirming Dr Horsfield's views As, however, in treating on the genus Thecla, I did not consider it advisable to separate T Rubi, although differing from the other species in the veins of the wings, so I shall not in the present genus separate Alsus generically from the rest, considering them too closely allied together to allow of such a step. My arrangement of the species is therefore as follows —

SECTION I —(Pithecops, Horsfield) First branch of the post or subcostal vein of the fore wings coalescing with the mediastinal or costal one, and subsequently again branching off from it P Alsus

SECTION II.—(Polyommatus proper.) First branch of the post or subcostal vein free, and extending to the costa of the fore wings; the other veins as in Thecla proper.

SUBSECTION I.—Hind wings without a sub-marginal row of fulvous spots on the under side.

A. Wings broad, hind wings rounded: females blue above, with broad dark margin to the fore wings. P. Argiolus.

B. Wings more triangular, hind pair more ovate. P. Cymon, Arion and Alcon.

SUBSECTION II.—Hind wings with a submarginal row of fulvous spots on the under side, comprising all the other species, which may be divided into groups according to the colour of the upper side of the wings in the opposite sexes.

DESCRIPTION OF PLATE XXXI.

INSECTS.—Fig. 1. Polyommatus Argiolus, male (the azure blue Butterfly). 2. The female. 3. Showing the under side.

Fig. 4. Polyommatus Alsus (the Bedford blue Butterfly), female. 5. The male. 6. Showing the under side. 7. The Caterpillar. 8. The Chrysalis.

Fig. 9. Polyommatus Acis (the Mazarine blue Butterfly), male. 10. The female. 11. Showing the under side.

PLANTS.—Fig. 12. Medicago denticulata (Reticulated Medicago). 13. Astragalus Alpinus (the Alpine milk vetch).

The English names of P. Argiolus and P. Alsus seem to require reformation: the former seems inappropriately styled the *azure* blue, which term does not at all describe its peculiar tint, whilst others of the genus, P. Adonis, for example, might truly be styled "azure blue." I should propose calling it the "light blue." The name of P. Alsus, "the Bedford blue," is still less descriptive, for neither of the sexes is blue at all, though in some individuals the male has a sort of purple gloss in certain lights. The "Bedford brown" would be far more intelligible. Godart, from whose figure I have taken the Caterpillar and Chrysalis, describes the larva of Alsus as feeding upon Astragalus cicer; but as this plant is not found in England I have figured an elegant British species instead. The Medicago, upon which it is probable that the larva of some of the Polyommati may feed, I have been induced to give a figure of from the singularity of its seed vessels.—H. N. H.

SPECIES 1.—POLYOMMATUS (PITHECOPS) ALSUS. THE BEDFORD BUTTERFLY.

Plate xxxi. fig. 4—8.

SYNONYMES.—*Hesperia Alsus*, Fabricius.
Papilio Alsus, Gmelin, Lewin, Lep. Pap. pl. 39, f. 3, 4, Donovan, Brit. Ins. 9, pl. 322, fig. 1.
Polyommatus Alsus, Stephens, Curtis, Duncan Brit. Butt. pl. 31, f. 5. Wood, Ind. Ent. t. 2, f. 12.
Nomiades Alsus, Hubner (Verz. bek. Schmett.)
Papilio minimus, Esper, Schaffer, Villers.
Papilio Pseudolus, Borkhausen.

This is the smallest of our British Butterflies, the expanse of the fore wings generally varying from $\frac{6}{8}$ to 1 inch. On the upper side the wings are of an obscure brown colour, with a slight blue gloss towards the base, especially in the males, of which sex I possess a specimen, in which at least half of the atoms of the disk of the wings are silvery blue: the female, on the contrary, is more obscure. The fringe of the wings is white. On the under side all the wings are of a light ash colour, with a slender black lunule at the extremity of the discoidal cell: half-way between this lunule and the hind margin of the fore wings is a transverse row of black, ocellated spots, with white sides, the two inner ones being more confluent; the hind wings have three or four similarly ocellated spots, irregularly placed in the basal half of the wings, beyond the middle of which is a waved row of seven or eight similar spots; and on the margin of these wings is a black spot, at a short distance from the anal angle, unnoticed either by Haworth or Stephens, and several obsolete brown spots. The number of the spots on the disk of the wings is, however, liable to variation.

This plain-coloured little butterfly, remarkable for the great delicacy of the markings on the under side of the wings, appears at the end of May and beginning of July, and occurs in a number of localities in different parts

In Octavo, No I, price 3s 6d,

ARCANA ENTOMOLOGICA;

OR,

ILLUSTRATIONS OF NEW, RARE, AND INTERESTING EXOTIC INSECTS

BY J O WESTWOOD, Esq., F.L.S &c

CONTAINING

FOUR COLOURED PLATES, WITH DESCRIPTIONS.

In foolscap 8vo, cloth lettered, price 6s 6d.,

MEMOIRS OF BRITISH FEMALE MISSIONARIES.

WITH A PRELIMINARY ESSAY

BY MISS THOMPSON,

ON THE IMPORTANCE OF FEMALE AGENCY IN EVANGELIZING PAGAN NATIONS

EMBELLISHED WITH A FRONTISPIECE ON STEEL,

REPRESENTING THE BURNING OF THE WIVES OF RUNJEET SINGH

No. XII.] [Price 2s. 6d.

BRITISH INSECTS AND THEIR TRANSFORMATIONS.

BRITISH BUTTERFLIES

AND

Their Transformations.

ARRANGED AND ILLUSTRATED IN A SERIES OF PLATES

BY H. N. HUMPHREYS, ESQ.

WITH CHARACTERS AND DESCRIPTIONS

BY J. O. WESTWOOD, ESQ., F.L.S.,
SEC. OF THE ENTOMOLOGICAL SOCIETY, ETC. ETC.

LONDON:
WILLIAM SMITH, 113, FLEET STREET.

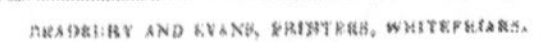

MDCCCXLI.

BRADBURY AND EVANS, PRINTERS, WHITEFRIARS.

WORKS PUBLISHED BY WILLIAM SMITH, 113, FLEET STREET

SMITH'S STANDARD LIBRARY.

In medium 8vo, uniform with Byron's Works, &c

Works already Published

ROKEBY BY SIR WALTER SCOTT 1*s* 2*d*

THE LADY OF THE LAKE 1*s*
THE LAY OF THE LAST MINSTREL 1*s*
MARMION 1*s* 2*d*
THE VISION OF DON RODERICK, BALLADS AND LYRICAL PIECES By SIR WALTER SCOTT 1*s*
THE BOROUGH By the REV G CRABBE 1*s* 4*d*
THE MUTINY OF THE BOUNTY 1*s* 4*d*
THE POETICAL WORKS OF H KIRKE WHITE 1*s*
THE POETICAL WORKS OF ROBERT BURNS 2*s* 6*d*
PAUL AND VIRGINIA, THE INDIAN COTTAGE, and ELIZABETH 1*s* 6*d*
MEMOIRS OF THE LIFE OF COLONEL HUTCHINSON, GOVERNOR OF NOTTINGHAM CASTLE during the Civil War By his Widow, Mrs LUCY HUTCHINSON 2*s* 6*d*
THOMSON'S SEASONS, AND CASTLE OF INDOLENCE 1*s*
LOCKE ON THE REASONABLENESS OF CHRISTIANITY 1*s*
GOLDSMITH'S POEMS AND PLAYS 1*s* 6*d*
———— VICAR OF WAKEFIELD 1*s*
———— CITIZEN OF THE WORLD 2*s* 6*d*
———— ESSAYS, &c 2*s*

, These four Numbers form the Miscellaneous Works of Oliver Goldsmith, and may be had bound together in One Volume, cloth-lettered, price 8*s*

KNICKERBOCKER'S HISTORY OF NEW YORK 2*s* 3*d*
NATURE AND ART By Mrs INCHBALD 10*d*
A SIMPLE STORY By Mrs INCHBALD 2*s*
ANSON'S VOYAGE ROUND THE WORLD 2*s* 6*d*
THE LIFE OF BENVENUTO CELLINI 3*s*
SCHILLER'S TRAGEDIES, THE PICCOLOMINI, and THE DEATH OF WALLENSTEIN 1*s* 8*d*
THE POETICAL WORKS OF GRAY & COLLINS 10*d*
HOME By Miss SEDGWICK 9*d*
THE LINWOODS By Miss SEDGWICK 2*s* 8*d*
THE LIFE AND OPINIONS OF TRISTRAM SHANDY By LAURENCE STERNE 3*s*
INCIDENTS OF TRAVEL IN EGYPT ARABIA PETRÆA AND THE HOLY LAND By J L STEPHENS, Esq 2*s* 6*d*
INCIDENTS OF TRAVEL IN THE RUSSIAN AND TURKISH EMPIRES By J L STEPHENS, Esq 2*s* 6*d*
BEATTIE'S POETICAL WORKS, and BLAIR'S GRAVE 1*s*
THE LIFE AND ADVENTURES OF PETER WILKINS, A CORNISH MAN 2*s* 4*d*
UNDINE A MINIATURE ROMANCE Translated from the German, by the REV THOMAS TRACY 9*d*
THE ILIAD OF HOMER Translated by ALEX POPE 3*s*
ROBIN HOOD, a Collection of all the ancient Poems, Songs, and Ballads now extant, relative to that celebrated English Outlaw, to which are prefixed, Historical Anecdotes of his Life Carefully revised, from RITSON 2*s* 6*d*
THE LIVES OF DONNE, WOTTON, HOOKER, HERBERT, AND SANDERSON Written by IZAAK WALTON 2*s* 6*d*
THE LIFE OF PETRARCH By Mrs DOBSON 3*s*
MILTON'S PARADISE LOST 1*s* 10*d*
———— PARADISE REGAINED, and MISCELLANEOUS POEMS 2*s*
RASSELAS By DR JOHNSON 9*d*
ESSAYS ON TASTE By the Rev ARCHIBALD ALISON, LL B 2*s* 6*d*
THE POETICAL WORKS OF JOHN KEATS 2*s*
THOMSON'S POEMS AND PLAYS, COMPLETE 5*s*
PICCIOLA A Tale from the French 1*s* 4*d*
THE POETICAL WORKS OF DR YOUNG, COMPLETE 3*s*

FOUR VOLUMES ARE NOW ARRANGED ON THE FOLLOWING PLAN —

ONE VOLUME OF "POETRY,"

CONTAINING

SCOTT'S LAY OF THE LAST MINSTREL
SCOTT'S LADY OF THE LAKE
SCOTT'S MARMION
CRABBE'S BOROUGH
THOMSON'S POETICAL WORKS
KIRKE WHITE'S POETICAL WORKS
BURNS'S POETICAL WORKS

A SECOND VOLUME OF "POETRY,"

CONTAINING

MILTON'S POETICAL WORKS
BEATTIE'S POETICAL WORKS
BLAIR'S POETICAL WORKS
GRAY'S POETICAL WORKS
COLLINS'S POETICAL WORKS
KEATS'S POETICAL WORKS
GOLDSMITH'S POETICAL WORKS

ONE VOLUME OF "FICTION,"

CONTAINING

NATURE AND ART By Mrs INCHBALD
HOME By MISS SEDGWICK.
KNICKERBOCKER'S HISTORY OF NEW YORK By WASHINGTON IRVING
PAUL AND VIRGINIA By ST PIERRE.
THE INDIAN COTTAGE By ST PIERRE
ELIZABETH By MADAME COTTIN.
THE VICAR OF WAKEFIELD By OLIVER GOLDSMITH
TRISTRAM SHANDY By LAURENCE STERNE

ONE VOLUME OF "VOYAGES AND TRAVELS,"

CONTAINING

ANSON'S VOYAGE ROUND THE WORLD
BLIGH'S MUTINY OF THE BOUNTY
STEPHENS'S TRAVELS IN EGYPT, ARABIA PETRÆA, AND THE HOLY LAND
STEPHENS'S TRAVELS IN GREECE, TURKEY, RUSSIA, AND POLAND

These Volumes may be had separately, very neatly bound in cloth, with the contents lettered on the back, price 10*s* 6*d* each

Those parties who have taken in the different works as they were published, and who wish to bind them according to the above arrangement, may be supplied with the title-pages gratis, through their booksellers, and with the cloth covers at 1*s* each

of the county South Creek, Norfolk, Brandon Warren, Suffolk, Dartmouth, near Andover, Hants, Birch and Darenth Woods Kent, and near Hertford, are mentioned by Mr Stephens near Darlington, in great abundance, by Mr J O Backhouse, near Amesbury, Wilts, by the Rev G T Rudd, near Newcastle, by Mr Wailes, near Durham, and in most of the northern counties of Scotland by Mr Duncan, and in the Isle of Wight, also between Woodstock and Enstone, near Cheltenham, Dover, &c by the Rev W T Bree

The caterpillar is green with yellow dorsal and lateral lines, it feeds upon Astragalus Cicer, according to Godart

---

SPECIES 2.—POLYOMMATUS ARGIOLUS. THE AZURE BLUE BUTTERFLY

Plate xxxi fig 1—3

Synonymes.—*Papilio Argiolus*, Linnæus Syst Nat 2, 790 Haworth Donovan British Insects, vol 11, pl 181 Lewin, Pap pl 36, fig 4 - 6
Lycæna Argiolus, Ochsenheimer, Leach, Samouelle
Polyommatus Argiolus, Latreille, Stephens, Curtis Wood Ind Entomol pl 2, fig 61 Duncan Brit Butt, pl 31, fig 1, 2
Agriades Argiolus, Hubner (Verz bek Schmett)
Papilio Acis Hubner, Pap, fig 272—4
Papilio Cleobis, Esper, Pap
Papilio Argus marginatus, De Geer, Gen Ins, 30, 3

This delicate butterfly measures from an inch and a sixth to an inch and a half in the expansion of its wings, which in the male are, on the upper side of a delicate light blue, with a tinge of a pinkish-blush the costa of the fore wings being still paler At the extremity of the wings in this sex, there is a narrow border of dark brown, which colour also extends at the tips of the principal veins of the wings into the fringe which is otherwise of a white colour, in the hind wings the fringe is entirely white, preceded by a very slender dark-brown line at the edge of the wings, the base of the wings is also darker On the under side, the wings are of a very delicate greyish-white, tinged with silvery blue, especially at the base of the hind wings, the fore wings are marked with a slender blackish transverse line at the extremity of the discoidal cell, beyond this are five or six black spots one placed a little in advance of the others, and nearer the fore margin of the wings, the others are more oblong and placed obliquely, that near the posterior angle being sometimes geminated, between this row of marks and the margin are several almost obsolete dusky crescents The hind wings beneath are marked with ten or twelve small black dots, placed irregularly, one of which is at the anal angle, besides which there appear traces of a submarginal row of dusky crescents above a row of dusky spots, there is also a very slender dusky line at the extremity of the discoidal cell

The female differs from the male in being generally of a smaller size, with the blue colour of the upper side of the wings somewhat paler, but is more particularly distinguished by having the extremity of the anterior and the entire outer margin of the fore wings marked with a broad black or dark-brown border The hind wings are also marked with a submarginal row of dark brown or black spots, which are sometimes so large as to be almost confluent, the costa of these wings is also dusky In other respects, as well as on the under side of the wings there is scarcely any difference between the two sexes The spots on the under side of the wings, as well as the dusky markings of the female, vary considerably in size, the former also differs in number in different individuals The caterpillar is pubescent, of a greenish-yellow colour, with a bright green line down the back, the head and legs being black It feeds on the buck thorn and holly

The chrysalis is smooth brown and green, with a dark dorsal line

This pretty species differs materially in its habits from its congeners frequenting gardens and plantations where the holly abounds.* It is by no means uncommon, although certainly local. Some years ago it appeared for two consecutive years in my garden, at Hammersmith, where some hollies had then recently been planted, but I have not since seen it. Epping Forest, near Ripley, near Dartford, and various parts of Norfolk, Suffolk, Hants, and Devonshire, are recorded as its localities by Mr Stephens. Not unfrequently near Newcastle, in places where hollies abound, and also in Castle Eden Dean, by Mr Duncan. The Rev W T Bree informs us it is common near Allesley in the early spring (as early as the middle of April), and that he has taken it in the Isle of Wight in the month of July. The middle of May and end of August are given as the times of its appearance by Haworth and Stephens, but the Rev W T Bree states that it seems to be only single brooded near Allesley; during the present season he has not, however, observed it in any of its usual localities near Coventry. In Loudon's Magazine of Natural History, Nos 21, 23, 24, 27, 30, 65, and 66, are various communications relative to this butterfly chiefly connected with the question as to whether it is a single or double-brooded species.

SPECIES 3.—POLYOMMATUS ACIS. THE MAZARINE BLUE BUTTERFLY.

Plate xxxi fig 9—11

SYNONYMES.—*Papilio Acis*, Wiener Verz, Fuest 1, pl 42 fig 8a—d.
Lycæna Acis Ochsenheimer.
Polyommatus Acis, Stephens, Curtis. Wood Ind Ent t 2 f 63. Duncan Brit Butt pl 31, fig 1.
Nomiades Acis Hubner (Verz bek Schmett.)
Papilio Cymon Lewin, Pap pl 38 f 6, 7. Haworth. Jermyn.
Lycæna Cymon, Leach, Samouelle.
Papilio Argiolus, Esper Schmett t 21, f 1. Hubner Pap p 56 f 267—9.
Papilio Semargus, Borkhausen.

This very distinct species differs from the two preceding in the complete diversity in the colour of the upper surface of the wings of the two sexes, being blue in the male and dark brown in the female. The expansion of the wings is rather more or less than an inch and a quarter. The upper side of the wings, in the male, is of a dark-purplish blue, the costa of the fore wings with a very thin edging of white. The outer margin in all the wings is narrow and dark brown, which colour runs up into the wing along the veins; the fringe of all the wings is white. Beneath, the wings are of a pale greyish-brown, the base being saturated with blue; there is a slender transverse dark line at the extremity of the discoidal cell of each wing, beyond which is a curved row of irregular sized black spots, margined with white rings, there being sometimes as many as seven such spots on each wing, that near the anal angle of the hind wings being minute and doubled; there are also sometimes one or two ocellated spots near the base, but the number of these spots is liable to considerable variation. All the wings have a very narrow outer marginal line of darker brown.

The female differs from the male in having the upper side of all the wings dark brown, sometimes with a slight purplish coloration towards the base in both pair of wings.

This rare species frequents chalky districts. The late Mr Haworth gave Yorkshire and Norfolk as its localities, and Miss Jermyn, Sherborne, &c, Dorsetshire. Various parts of Cambridge, Hampshire, and Windlesham-heath, Surrey, are mentioned by Mr Stephens. There are also some notices of this insect in the 31st and 32nd Numbers of Loudon's Magazine of Natural History, by the Rev W T Bree, who informs us that he once took it in Coleshill Park, Warwickshire, also near Hinkley, Leicestershire; other specimens have also been taken in Worcestershire.

---

* It is a more restless and high-flying insect than any of the other Polyommati, hovering and vapouring about the trees and bushes. Mr Bree also observes that it does not evince the same partiality for settling upon flowers and leaves of humble growth, as it does for settling on the leaves of the holly.

DESCRIPTION OF PLATE XXXII

INSECTS.—Fig 1 Polyommatus Arion (the large blue Butterfly) 2 The female 3 Showing the under side

Fig 4 Polyommatus Alcon (the Alcon blue Butterfly) 5 The female 6 Showing the under side

PLANTS.—Figs 7 and 8 Trifolium fragiferum (the strawberry-headed Trefoil)

P Arion is drawn from specimens in the cabinet of Mr Stephens P Alcon from the accurate figure of Hubner, as I was unable to procure a well authenticated British specimen. It is not ascertained what plants the larvæ of these two species feed upon, and the larvæ themselves are yet unknown It is probable, however, that the Caterpillar of P Arion feeds upon some species of Trefoil, and as the perfect insect is found in marshy meadows, I have grouped it with a plant of the singular Trifolium fragiferum which grows in such situations H N H

SPECIES 4.—POLYOMMATUS ALCON THE ALCON BLUE BUTTERFLY

Plate xxxii fig 4—6

SYNONYMS.—*Hesperia Alcon*, Fabricius
Papilio Alcon, Hubner, Pap pl 55, f 263. 4—5
Lycæna Alcon, Ochsenheimer, Schmett v Europa 1, p 7 No 5
Polyommatus Alcon Stephens, Wood, Ind Ent t 53, fig 16
Duncan, Brit Butt pl 32, fig 2
Argus Alcon Boisduval, Icon Hist Lepid, vol pl 13 f 1—3
Papilio Arcas Esper Schmett t 34 Suppl 10 no 1, 5
Papilio Diomedes Borkhausen,
Hesperia Arqiades, Fabricius ?

This species appears to be intermediate between P Acis and Arion The expansion of the fore wings is about an inch and a half The upper side of the wings in the male is of a shining violet-blue, the middle vein and the costal margin having a silvery-white tinge, the extremity of all the wings is ornamented with a rather broader blackish border, the fringe being white, the discoidal cell of the fore wings is sometimes closed by a thin blackish transverse streak The under side of all the wings is of a darkish ashy-grey colour, with the base suffused with blue, each is marked at the extremity of the discoidal cell with a black crescent bordered with white half-way between which and the extremity of the wings is a sinuous row of black spots, each encircled with yellowish-grey, between this row of spots and the outer margin of the wings is another series of black ocellated dots, succeeded close to the margin by a row of almost obsolete lunules. There are also two or three ocellated spots near the base of the wings The fringe is white above, but spotted with black beneath

The female differs in having the upper surface of the wings strongly suffused with black along the anterior and outer edges of the fore wings, and leaving a large patch along the inner edge of the fore wings, and the disk of the hinder wings saturated with blue, in addition to this there is a transverse black line at the end of the discoidal cell, and a row of black spots between the latter and the outer margin of the wing, the fringe of the female is of a reddish-grey colour The hind wings also bear a nearly obsolete series of black spots between the middle and the hind margin

This species, which is abundant in some parts of France and other Continental districts, has been introduced by Mr Stephens into the British lists, on the authority of a specimen formerly in Mr Haworth's collection captured by the late Mr Jones, in Buckinghamshire many years since Mr Stephens, however, suggests that Mr Haworth's specimen may prove to be only an extraordinary variety of Pol Arion

SPECIES 5.—POLYOMMATUS ARION. THE LARGE BLUE BUTTERFLY.

Plate xxxii. fig. 1—5.

SYNONYMES.—*Papilio Arion*, Linnæus, Faun. Suec. 1073. Haworth. Lewin, Pap. pl. 37. f. 1—2. Donovan, Brit. Ins. v. 6, p. 184. fig. Hubner, Schmett. pl. 54. fig. 254—6. *Lycæna Arion*, Ochsenheimer. Leach. *Polyommatus Arion*, Latreille, Stephens, Curtis, Wood, Index Ent. t. 3. fig. 64, ♂ ♀. Duncan, Brit. Butt. pl. 32, fig. 1. *Nomiades Arion*, Hubner (Verz. bek. Schmett.)

This fine and very rare species generally measures somewhat more than an inch and a half in the expanse of its wings, which are of a rather dark purplish-blue in the males, with the anterior or costal margin pale brown; but the outer margin in all the wings is rather broadly black; in addition to this, the males may be distinguished from those of every other indigenous species by having a black crescent at the extremity of the discoidal cell, and five black oval spots between it and the dark border on the upper side of the fore wings; the hind wings have also several black oval dots beyond the centre, and a submarginal row of black spots ocellated with blue; the fringe is white. Beneath, the ground colour of the wings is ashy-grey, rather darker in its tone than in P. Acis, strongly suffused with shining blue atoms, at the base especially, on the hind wings. The fore wings have one circular and one kidney-shaped black spot, ocellated with whitish in the discoidal cell, beyond which is a very curved row of large black spots similarly ocellated; parallel with the outer margin of the fore wings, are two rows of black spots, separated from each other by whitish atoms, and the extreme margin of the wings is also black, and the fringe, which is white, is marked at the tips of the longitudinal veins with black spots. Each of the hind wings is marked with about twenty-six black spots, which (with the exception of those which are nearest the margin of the wing) are ocellated with whitish; three of these spots form a curve near the base of the wings, and are followed by a curved short transverse line at the extremity of the discoidal cell, this is succeeded by an irregularly curved row of eight ocellated spots; beyond this, are two rows of spots, parallel with the posterior margin, and the fringe is marked in the same manner as in the fore wings.

The female is distinguished by having the wings more suffused with dark brown, and the spots on the disk of the fore wings are larger and longer than in the males; the spots are also occasionally more numerous, and the edges of the wings with a broader dark margin. There are, however, several varieties described, in which the number and size of the spots varies considerably, and Mr. Stephens mentions one variety in which the wings are almost immaculate above.

Mr. Haworth received this species from Dr. Abbott, who took it near Bedford, in the Monse's Pasture, where Mr. Dale again took it in 1819. It is also recorded as having been taken on Dover Cliffs, Marlborough Downs, and on the hills near Bath; also on commons near Broomham, Bedfordshire, near Winchester, and on bramble blossoms in some parts of North Wales. The species has also been recently taken by Mr. Queckett in some profusion, as well as by Mr. Bree's son, in the middle of July, 1837-8 and 9, at Barnewell Wolde, near Oundle, Northamptonshire, where Mr. Bree himself also found it on the 4th of July, 1840.

DESCRIPTION OF PLATE XXXIII.

INSECTS.—Fig. 1. Polyommatus Adonis (the Clifden, or azure blue B.) 2. The female. 3. Showing the under side.

" Fig. 4. Polyommatus Corydon (the Chalk hill blue B.) 5. The female. 6. Showing the underside. 7. The Caterpillar. 8. The chrysalis.

PLANTS.—Figs. 9 and 10. *Trifolium stellatum* (the starry headed Trefoil.)

All these insects are figured from beautiful specimens in the cabinet of Mr. Stephens. The Caterpillar of P. Corydon I have represented feeding upon Trifolium stellatum, which I have selected for the singular appearance of the seed-vessels. The Caterpillar is from Hubner. H. N. H.

SPECIES 6 —POLYOMMATUS CORYDON THE CHALK-HILL BLUE BUTTERFLY

Plate xxxiii fig 1—8

SYNONYMES —*Hesperia Corydon*, Fabricius, Hubner, Pap pl 59 286—8 Lewin's Pap pl 36, figs 1, 2, 3 Donovan, Brit Insects pl 231, f 1, male Esper Schmett t 33, fig 4
Polyommatus Corydon, Latreille, Stephens, Jermyn, Curtis Wood Ind Ent pl 2 fig 65 Dun in Brit But pl 32, fig 3
Agriades Corydon, Hubner (Verz bek Schmett)
Papilio Tiphys, Esper Pap pl 21 cont 1, fig 4 (Female)
Papilio Calæthys, Jermyn, 2d Edit p 109 (variety)

This species varies in the expansion of its wings from an inch and a third to more than an inch and a half The male has the upper surface of the wings of a very light silvery blue, with the outer margin and veins dusky close to the outer margin is a row of black spots, which are almost suffused with dark margin of the fore wings but are more distinct in the hind pair, two at the anal angle being smaller and close together, these spots are more or less annulated with silvery white The fore wings in this sex are of a greyish white on the under side, with five rows of spots, four towards the base in pairs, one larger, and one smaller in each, then a transverse nearly straight row of four spots of which the inner one is doubled, succeeded by a curved row of four spots towards the costa then two submarginal rows of dots, the inner ones being the largest, forming an interrupted bar, and the outer ones rounded and subocellated, the tips of the veins, and of the cilia opposite to the veins, are marked with dark spots, the disc of the hind wings is of a pale-greyish brown hue, the base strongly saturated with greenish blue, each marked with about twenty blackish ocellated spots, an almost blind white spot at the extremity of the discoidal cell, the space beyond the two middle spots in the outer curved series is also white Seven or eight of the terminal spots are ocellated, each being preceded by an angular black mark, and a small patch of orange colour, the extreme edge of the wing is also blackish, and the fringe is white

The female differs from the male in having the upper surface of the wings of a brown colour, with a small paler spot in the middle of each, that in the fore wings having a black pupil, moreover, there is a submarginal row of ocelli having the pupil black, surrounded by a whitish iris, the upper part in the hind wings being orange, these ocelli are also sometimes preceded by a row of almost obsolete pale lunules, in some specimens, however, the appearance of these ocellated spots is almost lost, on the underside the ground colour of all the wings is considerably darker than in the males, and the ocelli are much more distinct, they are, however, similar in their number and situation to those of the male, but the fringe is more strongly marked alternately with brown

There are a number of varieties in our cabinets resulting from the greater or less distinctness of the ocelli and the greater suffusion of brown over the wings of the male One of these varieties having the wings "above, brown with a blue disc, and a whitish discoidal dot with a black pupil, beneath, the posterior wings have a discoidal, white, cinctured crescent, with a waved band of seven undulated spots towards the hinder margin," constitutes the Polyommatus Calæthys, of the second edition of Miss Jermyn's Butterfly Collector's Vade mecum

The caterpillar is green, with yellow dorsal and lateral lines It is stated to feed upon the wild thyme The perfect insect appears in July It is local in respect to the districts in which it is found, especially frequenting chalky places In such places it is, however, very abundant From Dover, along the southern coast near Shoreham, Newport, in the Isle of Wight, and near Darenth Wood in Kent, various parts also of Suffolk Oxfordshire, Cambridgeshire, are recorded as its localities It is also "very abundant on the hills above Prestbury, near Cheltenham, and near Winchester A single specimen was also taken a few years ago near Knowle, Warwickshire," as we are informed by the Rev W T Bree

SPECIES 7.—POLYOMMATUS ADONIS. THE CLIFDEN BLUE BUTTERFLY.

Plate xxxii fig 1—3

SYNONYMS.—*Hesperia Adonis*, Linnæus.
Papilio Adonis Lewin, Pap pl 38 fig 1—3; Haworth.
Lycæna Adonis Ochsenheimer; Leach; Samouelle.
Polyommatus Adonis Stephens; Curtis; Wood, Ind Ent pl 2, f 66; Duncan, Brit Butt pl 3, fig 1—2.
Papilio Argus Donovan, Brit Ins pl 143 fig 1 female (not the upper figures which belong to P Alexis and not the Pap Argus of Linnæus).
Papilio Ceronus, Hubner, Pap pl 295—297.
Papilio Bellargus Esper; Villers; Muller.

This, the most splendid of all the British blues, varies from $1\frac{1}{4}$ to $1\frac{1}{2}$ inches in the expanse of the wings, which in the males are of a most lovely, shining, silvery, azure blue; the costa of the fore wings rather more silvery, and the outer margin of the wings with a slender dark line, the fringe white, with small brown patches at equal distances. On the under side the ground colour of the wings is darker than in the corresponding sex of P Corydon, and the ocelli are more strongly marked, although nearly similar in their situation; there is, however, only a remote spot preceding the dot at the end of the discoidal cell of the fore wings, and the succeeding series of spots is more continuous, the fifth from the costa not being thrown so much forward as to break the curve, as it is in P Corydon. The ocelli and other spots on the under side of the hind wings are, however, almost exactly placed as in that species, and they are also similarly coloured.

The female has the upper surface of the body and wings of a dark brown colour, the disc towards the base being sometimes saturated with blue; there is a small black spot at the extremity of the discoidal cell in each wing, and in the hind wings there is a submarginal row of ocellated black spots, the inner part of the iris of each being marked with an orange curve, the ocelli towards the outer angle being almost obliterated; some specimens also, have the rudiments of a series of fulvous arches near the outer margin, the fringe is brownish white, interrupted with brown spots; on the under side the ground colour of the wings, as in P Corydon, is darker than in the males, and the ocelli larger and more conspicuously ocellated with whitish, although similar in their situation.

The position, size and number of the ocelli on the under side are liable to some variation; and I possess several specimens in which the opposite sides do not exactly correspond with each other in these particulars: the white blotch on the hind wings and the orange submarginal spots are also sometimes almost obliterated.

The caterpillar is described by Fabricius as being green, with dorsal rows of fulvous spots. The perfect insect appears to be double brooded, the first specimens appearing at the end of May, and the others at the middle of August. It occurs in various parts of the southern counties of England, especially in chalky districts, in some profusion. It also occurs in some parts of Suffolk. As it is by far the most lovely of the British blues, it used to be much sought after by the Spitalfields collectors, who, as Mr Haworth states, made distant pedestrian excursions for the sole purpose of procuring its charming insects to decorate their pictures with —a picture, consisting of numerous and beautiful Lepidoptera ornamentally and regularly disposed, having been the ultimate object of these assiduous people in the science of Entomology. These pictures were of various shapes and sizes, and Mr Haworth mentions having seen some which contained at least five hundred specimens. Such was the custom some twenty-five years ago, and it is this class of persons whose feelings Crabbe thus records in his 'Borough'—

> 'There is my friend the weaver; strong desires
> Reign in his breast; tis beauty he admires.
> See to the shady grove he wings his way
> And feels in hope the rapture of the day—

Eager he looks, and soon to glad his eyes
From the sweet bower by nature formed arise
Bright troops of virgin moths and fresh born butterflies—
—He fears no bailiff's wrath, no baron's blame
His is untaxed and undisputed game "

Indeed so strong is the "fancy," as it is termed, with some of these laborious collectors that I have known some, who, after toiling at their weaving machines all the week have started at ten o'clock on Saturday night, in order to arrive at Darenth and Birch Wood by daybreak, so as to collect the twilight-flying moths Daniel Bydder one of the most industrious of these collectors, and who was employed by Dr Leach to collect for him in the New Forest (where he discovered Platypus cylindrus and Cicada anglica), was, I believe the first of the Spitalfields collectors who attempted to arrange his insects scientifically and now, following the example of the Entomological Society, they have formed themselves into a society of "Practical Entomologists," and have a well-arranged collection, meeting at regular intervals, in order to communicate to each other the result of their captures

DESCRIPTION OF PLATE XXXIV

INSECTS —Fig 1 Polyommatus Argus, (the silver studded blue Butterfly) 2 The female 3 A common variety of the female 4 Showing the under side 5 The Caterpillar 6 The Chrysalis

" Fig 7 Polyommatus Alexis (the common blue Butterfly) 8 The female 9 Showing the under side 10 A variety of the female 11 An Hermaphrodite variety having the wings of a female on one side and of a male on the other 12 The Caterpillar 13 The Chrysalis

PLANTS —Fig 14 Cytisus Scoparius (the common broom)

" Fig 15 Medicago sativa (the cultivated lucerne)

The insects on the present plates are from specimens in the British Museum, where are three other Hermaphrodite specimens of P Alexis but none of any of the other species The caterpillars are both from Godart Hubner has given a very different figure of the larva of Alexis but as Godart minutely describes the rearing of several, I have preferred his figure He describes the larva of P Alexis as feeding upon the cultivated lucerne, and those of P Argus upon the common broom H N H

SPECIES 8 —POLYOMMATUS ALEXIS THE COMMON BLUE BUTTERFLY

Plate xxxiv fig 7—12

SYNONYMES —*Papilio Alexis*, Wiener Verzeichniss, p 181, Hubner, Pap pl 60 fig 292

Polyommatus Alexis, Latreille Stephens, Curtis, Wood Ind Ent pl 3, fig 69, m and f Duncan, Brit Butt title page

Papilio Icarus, Villars, Haworth Lewin, Pap pl 38, fig 4, 5, 8, Esper Schmett t 32, fig 4, m

Papilio Argus, Wilks, pl 119, Donovan, Brit Ins pl 143, upper figures, Harris Aurelian, pl 39, fig g—i

Lycæna Dorylas Leach Samouelle but not of Hubner

Papilio Hyacinthus, Lewin Pap 37 fig 4 5, 6 nec Fabricius Haworth (variety)

Polyommatus Labienus, Jermyn (variety)

Polyommatus Thestylis Jermyn (variety)

Polyommatus Icaron, Jermyn (variety)

Polyommatus dubius Kirby MSS (variety)

This, one of the most abundant of our native butterflies, varies in the expanse of its wings from less than an inch to nearly an inch and a half The upper surface of the wings in the males is of a fine silky lilac-blue, the anterior margin of the fore wings being edged with white, the outer edge of all the wings with a slender dark line and the fringe white The body is clothed with long whitish blue silken hairs The under side of the wings is also very similar in its marking to the two preceding species, but the ground colour of the wings is rather paler There is an ocellated spot in the middle of the discoidal cell, with another, more indistinct, beneath it, which is sometimes connected with the innermost ocellus of the series between the extremity of the discoidal cell and the outer margin of the wing, the base of the hind wings is strongly glossed with shining bluish-green atoms, and the sub marginal row of fulvous markings on the hind wings is very distinct the marks at the anal angle being duplicated In the centre of the hind wings is a triangular white spot, generally with a black dot in the centre, preceded, towards the base, by four ocelli, placed obliquely, and between the middle ocelli of the row beyond the centre of the wing and the orange spots is a white patch, there is also a slender black marginal line, and the fringe is white

The female differs in having the upper side of the wings brown, the disc more or less suffused with blue, there is also a submarginal row of fulvous spots, which are sometimes obsolete in the fore wings, in the hind wings they are preceded by black lunules and succeeded by black sub-ocellated spots On the underside the ground colour of the wings is browner than in the males, and the ocelli larger and more distinct The base in these wings is also less strongly tinged with green The fringe in this sex is rather darker than in the male, especially at the base, but not spotted, by which it is at once known from the female of P Adonis Varieties occur in this as in the preceding species, in which the number and size of the ocelli beneath, and markings on the upper side, are more or less obliterated I possess indeed some specimens in which the opposite sides are not alike in these respects

One of these varieties, which Mr Haworth thought might be a hybrid between Adonis and Alexis, has the two spots towards the base of the fore wings, on the under side, obsolete, and the upper side of the wings of the female more strongly saturated with blue These form the species P Hyacinthus of Lewin and Haworth

Others again, of very small size (not expanding more than 10½ lines), have the upper side of the wings of a very pale lilac-blue, and the spots on the under side very small and pale the interior spot at the base of the fore wings obsolete, only five spots in the curved row beyond the middle of the discoidal cell, and the fulvous lunules almost obsolete, the two basal spots on the costa of the hind wings large and black I have made this description from Mr Kirby's original specimen on which the Polyommatus Labienus was proposed

Polyommatus Thestylis of Jermyn is formed upon large female specimens of this species, in which the blue of the upper surface of the wings is much more extended than in ordinary individuals, "the anterior wings beneath with a large kidney-shaped blackish spot cinctured obscurely with white, the concave side turned towards the interior margin the posterior wings with the spot next the costal margin kidney-shaped, the concave side turned towards the disc, the number of ocelli in all the wings varies considerably, and the kidney-shaped spot is sometimes interrupted"

Polyommatus Lacon of Jermyn is another variety, in which the disc of the wings beneath is only marked with a triangular spot, the hind margin of the anterior with a few indistinct dusky marks, and of the posterior with a fulvous band terminated internally with a series of black wedge-shaped spots, and externally with black dots on a white ground'

Mr Stephens also adds that some specimens even differ in form from the rest, some of the females having the anterior wings very much rounded at the tip, whilst in others they are somewhat acute In some females also the disc of the wings on the upper side is entirely brown, whilst in others it is nearly as blue as in the males, with a black discoidal spot

As some of the preceding varieties appear to be constant in certain localities, Mr Stephens informs me that he has but little doubt that they in fact constitute distinct species, such is particularly the case with certain individuals of the males, which have the wings very transparent, and of a more silvery blue, and the females very blue, with very distinct red lunules adjoining the black submarginal and distinct ocelli

(To be completed in Fourteen Nos.)

No. XIII.] [Price 2s. 6d.

BRITISH INSECTS AND THEIR TRANSFORMATIONS.

BRITISH BUTTERFLIES

AND

Their Transformations.

ARRANGED AND ILLUSTRATED IN A SERIES OF PLATES

BY H. N. HUMPHREYS, ESQ.

WITH CHARACTERS AND DESCRIPTIONS

BY J. O. WESTWOOD, ESQ., F.L.S.,
SEC. OF THE ENTOMOLOGICAL SOCIETY, ETC. ETC.

LONDON:
WILLIAM SMITH, 113, FLEET STREET.

MDCCCXLI.

This species also appears to be subject to gynandromorphism to a greater degree than any other of our butterflies, although this is probably owing to its being a more abundant species. Several instances of this are contained in the British Museum Cabinet, one of which is represented in our figure 11; other instances are recorded in the Annales de la Société Entomologique de France, The Field Naturalist, &c.

The caterpillar is slightly pubescent, and of a bright green colour, with a dark dorsal line, adjoining to which are rows of yellow spots. It is found at the end of April and of July, and feeds upon different grasses. The wild liquorice and wild strawberry are also mentioned by Mr Stephens as its food.

This common insect seems to be distributed over all parts of the kingdom, and is double-brooded, the first appearing about the end of May (but later in the northern parts of the country), and the second in August. It frequents meadows, grassy places at the sides of lanes, and pasture-lands. Mr Knapp thus describes some of its habits:—"We have few more zealous and pugnacious insects than this little elegant butterfly, noted and admired by all. When fully animated, it will not suffer any of its tribe to cross its path, or approach the flower on which it sits, with impunity; even the large admirable Atalanta at these times it will assail and drive away. Constant warfare is also kept up between it and the small copper butterfly; and whenever these diminutive creatures come near each other, they dart into action, and continue buffeting one another about till one retires from the contest, when the victor returns in triumph to the station he had left. Should the enemy again advance, the combat is renewed; but should a cloud obscure the sun, or a breeze chill the air, their ardour becomes abated, and contention ceases. The pugnacious disposition of the Argus butterfly soon deprives it of much of its beauty; and unless captured soon after its birth, we find the margins of its wings torn and jagged, the elegant blue plumage rubbed from the wings, and the creature become dark and shabby."—*Journal of a Naturalist*, p. 277.

SPECIES 9.—POLYOMMATUS ARGUS. THE SILVER-STUDDED BLUE BUTTERFLY.

Plate xxxv. fig. 1—6.

SYNONYMES.—*Papilio Argus*, Linnæus, Faun. Suec. 1074; Lewin, Pap. pl. 39, fig. 5—7. Haworth (not P. Argus of Donovan, vol. 4, pl. 143 ♂, which is the male of P. Alexis).
Hesperia Argus, Fabricius.
Lycæna Argus, Leach, Ochsenh., Hubner Pap. tab. 64, f. 316, ♂, 317, 318 ♀.
Polyommatus Argus, Stephens. Duncan Brit. But. pl. 33, fig. 3. Wood, Ind. Ent. t. 2, fig. 71.
Lycænides Argus, Hubner (Verz. bek. Schmett.).
Papilio Idas, Linnæus, Faun. Suec. 1075 (female), (P. Argus β, Linn. S. N. 2, 790), not P. Idas of Lewin, Donovan, and Haworth, which is P. Agestis.
Hesperia Aegon, Fabricius (variety).
Papilio Argyrognomon, Borkhausen (variety).
Papilio Argyades, Esper Papil. 1, pl. 101, cont. 56, fig. 6 (variety).
Polyommatus Aleppe, Kirby's manuscripts, in Mus. Ent. Soc. Lond. (variety).
Polyommatus maritimus, Haworth's manuscripts (variety).
Pap. Leodorus, Esper. Pap. 1, pl. 80, cont. 30, fig. 1, 2 (variety).

This pretty butterfly generally measures about an inch and a quarter in the expansion of the wings, which have the upper surface, in the male, of a fine, deep, lilacy blue, with the front margin of the anterior pair silvery white, of which colour also are the hairs on the wings, especially in the hind pair. The apical margin of all the wings on this side is broad and black, the dark colour slightly ascending along the veins into the disk of the wing, and in the hind wings assuming the appearance of oval, marginal spots. The cilia, both above and below, are white, a very slight black spot at the extremity of each of the veins being alone visible at the base of the fringe; the body above is clothed with silvery and blue hairs, the eyes are margined with white, and the antennæ are black, with white rings, the upper side of the club black, and the lower fine orange. Beneath, the wings in this sex are of a pale greyish colour, the base being saturated with blue; at the extremity of the discoidal cell of the fore wings is an oval, black ocellus, edged with white; this is succeeded by a curved row of six similar ocelli, varying in their form and size, the innermost of these spots is often doubled then follow two rows of dark spots, the inner row formed of arched spots, and the outer one of round, smaller ones; the space preceding the former is whiter than the rest of the wing, and the space between the two rows of spots is often coloured with orange; the margin of all the wings is slender and black, and triangularly dilated at the extremity of the veins. The hind wings have more numerous ocelli, namely, three small, round ones near the base, one transverse near the middle, a much curved and irregular row of eight beyond the middle beyond which the wing is whiter than in the other parts; then follows a curved row of eight black arches, and a fulvous band, on which are about the same number of round black spots, most of which are adorned with silvery scales.

The females are larger than the males, and have the upper surface of the wings of a dull warm, brown colour, darker at the base, and near the extremity of the wings is a series of fulvous, arched spots, occasionally more obsolete on the fore wings; along the margin of the hind wings is also occasionally a very slender, dull white interrupted streak. The ciliæ are also dusky, especially at the base. Beneath the ground colour of the wings is darker grey, or brownish ashy, which throws the white ocelli and other markings beyond the middle curved row of spots into stronger contrast. Moreover, the submarginal orange band is brighter coloured, the dark marking by which it is edged being more distinct. The female often differs by having the disk of the wings on the upper side more or less (and especially in the hind wings, as in our figure 3) suffused with blue.

Some striking varieties of this species have been observed. In one, captured by the late Mr. Hatchett at Coombe Wood, the upper surface of all the wings is of a pale fulvous tawny colour like that of Hipparchia Pamphilus. In another, taken by the late Mr. Haworth, in salt marshes, near Holt, Norfolk, and thence named by him P. maritimus, the ocelli on the disk of the under side of the wings are elongated into those on the middle of the wing, being almost confluent with the following row of spots. To a specimen of this variety in the cabinet of the Entomological Society of London, is attached the manuscript name of Alcippe of Kirby; but Mr. Stephens applies that name to another, and apparently very distinct variety, of smaller size, having "the wings *narrower*, blue above, with a broad, black margin to all the wings, the under side of the male of a deep greyish or drab colour, the ocelli very distinct in the female, and the oblique series on the posterior wing consisting of four."

The caterpillar is described as being of a dull green colour, with whitish tubercles, and a blackish head and legs; a line down the back and sides, oblique marks on the latter, of a dark red colour, bordered with white. It feeds on broom, sainfoin, and other kinds of Trifolium and allied genera. The chrysalis is at first green, and afterwards brown.

This species frequents lanes, marshy commons, damp fields, &c., about the middle of July, not appearing to be attached to chalky districts. Although not apparently found in the north of England it is sufficiently common in various parts of the south; Coombe and Darenth Woods, Ripley Green, Wood Hay Common, Hants, Parley Heath Dorset, Coleshill Heath Warwickshire (as we learn from the Reverend W. T. Bree), and other localities are recorded by preceding authors.

DESCRIPTION OF PLATE XXXI

INSECTS.—Fig 1 Polyommatus Eros 2 Showing the under side

„ Fig 3 Polyommatus Dorylas 4 The female 5 Showing the under side

PLANTS.—Figs 6 and 7 Hippocrepis comosa (the horse shoe Vetch)

I have not met with any British specimens of either of these species, those marked Dorylas in many collections being merely varieties of Adonis, and I have seen small varieties of Argus marked Eros It seems doubtful whether the true species has ever been found in this country, I therefore give the figures of Hubner to enable collectors to compare specimens, which they may suppose to be either of these species H N H

SPECIES 10.—POLYOMMATUS EROS THE PALE BLUE BUTTERFLY

Plate xxxi fig 1—2

SYNONYMES —*Lycæna Eros* Ochsenheimer, i, p 26 *Polyommatus Eros*, Jermyn, Steph Ill Haust i, p 93 Wood Ind Ent t 3, fig 70, ♂, ♀ | *Argus Eros* Boisduval, Icones Hist Lep pl 14 fig 4—6 *Papilio Tithonus*, Hubner Lep p 103 fig 555, 556, Haworth in Ent Trans 1 334

The expansion of the wings of this doubtful British species is rather less than an inch and a quarter The upper surface of the wings in the males is of a pale azurine blue, with a silvery or greenish tinge, which is not found in the allied species, P Dorylas The fringe is white, and separated from the ground colour of the wings by a black border, which is broader than in Dorylas, and which extends slightly upwards along the tips of the veins especially in the under wings which thus assume the appearance of having the margin spotted with black

The under side of the wings has considerable resemblance with Alexis in the ground colour and arrangement of the ocelli and markings, the ground colour being brownish grey, with the base saturated with greenish blue The fringe is separated from the ground colour by a slender black edge preceded by marginal lunules of pale yellowish orange, within each of which is a black arch, and on the opposite side a small black spot These fulvous markings are more indistinct on the upper wings their black anterior part is, on the contrary, more strongly marked Between the mark at the extremity of the discoidal cell of the fore wings (which is nearly the same as in Alexis) and the fulvous lunules, is a range of ocellated spots, the discoidal spot is preceded by an ocellated spot (Boisduval figures two such basal spots) in the fore wings, and in the hind wings by a row of four (or three) similar ocelli, placed upon the greenish part of the base, as in Alexis The under side of the breast and legs are of a greenish grey, and that of the abdomen white, the thorax and the under side of the abdomen are blue

The female differs from the male almost in the same manner as in Alexis, some specimens being entirely brown, and others saturated with blue from the base to beyond the middle, some again have only some blue atoms at the base, and along the inner edge of the fore wings The ground colour of the wings is rather black than brown, with a central black lunule The extremity of the hind wings is ordinarily marked with a row of lunules, fulvous within, black in the middle, and edged externally with white The under side scarcely differs from that of the male, except that the fulvous spots are rather brighter coloured

On the Continent this is a rare species, being found on the Alps of Valois, Tyrol, and France Mount Cenis and the environs of Digne and of Gap have been recorded as its localities It is therefore doubtful whether Messrs Haworth and Stephens, who have introduced this species into the English catalogues, may not have mistaken some pale variety of Alexis for it From that species it is at once known by the peculiar pale upper surface of the wings of the male, with a dark border, and from the female, by the much darker upper surface

Mr Haworth's specimen is supposed to have been taken in Kent. Mr Stephens gives his specimen as a variety, describing it thus —"Colour of the upper surface rather more intense, the inferior ocellated, nearly as in Alexis, but destitute of a fulvous, marginal fascia, in lieu of which it has a series of ocelli with minute black irides cinctured with white, faintly tinted with yellowish towards the inner side. This variety is probably synonymous with P. Labienus, of the first edition of the Butterfly Collector's Vade Mecum, unless var. γ of the preceding insect [Alexis] be the kind intended." The description which I have given of Mr Kirby's original specimen of P. Labienus in p. 108, will at once show that that supposed species has no connexion with the true Eros. Mr Stephens' specimen was taken in July 1826, in a grassy lane, near Ripley, Surrey.

SPECIES 11.—POLYOMMATUS DORYLAS. THE AZURINE BLUE BUTTERFLY.

Plate xxxv. fig. 3—5.

SYNONYMES.—*Papilio Dorylas* Fabricius. Wien Verz., Hubner Pap. pl. 67, fig. 289—291.
Lycæna Dorylas, Ochsenheimer.
Polyommatus Dorylas? Stephens Ill. Haust. I. p. 90. Wood, Ind. Ent. tab. 2. fig. 67.
Argus Dorylas, Boisduval, Icon. Hist. Lep. pl. 14, fig. 1—3.
Papilio Hylas, Esper. Schmetterl. pl. 45. Suppl. 21.
Papilio Thetis, Esper (female), pl. 55, cont. 5, fig. 1.
Papilio Golgus, Hubner, Europ. Schmett. (variety.)
L'azuré Ernst. Papil. d'Europe, pl. 83, Suppl. 2. pl. 4. fig. 32.

The expansion of the wings of this doubtful British species is rather more than an inch and a quarter. The upper surface of the wings of the male is of a bright azurine blue, nearly like Adonis, with the fringe and the anterior margin white, the fringe is preceded by a slender black margin, which extends a little along the veins of the wings, especially in the anterior pair. The under side is of an ashy-grey colour, slightly shaded with blue at the base, the discoidal cell of the fore wings is not marked with an ocellus in the centre, but is terminated by a curved, black spot, margined with white, beyond this is an irregular, curved row of six ocellated, black spots, succeeded by a row of fulvous arched spots, forming the inner edge of the pale margin of the wings. The under surface of the hind wings is marked with two, three, or four ocelli at the base, the extremity of the discoidal cell is occupied by a white spot, destitute of any black marks, and is succeeded by a curved row of ocelli, the middle ones of which are placed near or upon a patch of white, the extremity of the wings being marked with a row of fulvous crescents, preceded by black arches, and marked on the outside with black spots. The under side of the breast and the feet are of a bluish colour, and of the abdomen white. The upper side of the latter and of the thorax is blue, the antennæ are ringed with white.

The upper side of the female is of a uniform brown colour, with a marginal row of fulvous spots, sometimes on the hind wings alone, but occasionally on all the wings. The under side is of a reddish grey, not saturated with blue at the base, with similar spots to the male, except that the fulvous markings are brighter.

This species differs from P. Adonis in the fringe being unspotted, the fore wings beneath have not the basal ocellus, and the hind wings have the white spot at the extremity of the discoidal cell unspotted with black.

Boisduval gives the Alps, Pyrenees, and some parts of Hungary and Germany as the habitats of this species. Ochsenheimer gives Lewin's plate 38, figs. 1 and 2 as identical with this species, but Mr Stephens regards them as representing P. Adonis. The specimens also which were regarded as identical with this insect in Mrs Jermyn's Vade Mecum, 2nd edition, Mr Stephens further considers as possibly identical with his variety γ of P. Adonis, but that variety has the fulvous band on the hind margin of all the wings obliterated. The individuals still preserved in Mr Kirby's collection, presented to the Entomological Society, appear to me rather to be fine specimens of P Alexis, all having two spots at the base of the fore wings on the under side Mr Stephens states that these supposed specimens of Dorylas were taken in company with P Adonis in June 1812, but have not been met with afterwards

DESCRIPTION OF PLATE XXXVI

INSECTS —Fig 1 Polyommatus Icarius 2 The female 3 Showing the under side 4 The Caterpillar of P Icarius of Esper
,, Fig 5 Polyommatus Agestis (the brown Argus Butterfly) 6 The female 7 Showing the under side
PLANTS —Fig 8 Fragaria vesca (the wild strawberry)

The insect figured in this plate No 1—3, is the P Icarius of British collections, but is certainly not identical with the P Icarius of Esper which is the P Amandus of Hubner The present figure is from the collection of Mr Stephens, and appears but a variety of Alexis The caterpillar is that figured by Esper, which he has represented feeding upon the wild strawberry H N H

SPECIES 12 —POLYOMMATUS ICARIUS THE BLACK BORDERED BLUE BUTTERFLY

Plate XXXVI fig 1—4

SYNONYMS —*Papilio Icarius*, Esper Schmetterl part 1, p 35, tab 99, 54, fig 1 Haworth

Polyommatus Icarius, Jermyn, Stephens, Ill Haust 1, p 91 Wood, Ind Ent t 3, fig 68, ♂, ♀

Argus Icarius, Boisduval, Icones Hist Lep p 60, pl 12, fig 1—3

Lycæna Icarius, Ochsenheimer

Zephyrus Icarius, Dalm Trans Acad Stockholm, 1816 sp 19

Papilio Amandus Hubner, Pap tab 59 fig 283—285, Haworth (Ent Trans p 334)

Polyommatus Agathon, Godart, Encycl Meth t 9, p 69

Agriades Icarius, Hubner (Verz bek Schmett)

This is another doubtful British species, introduced by Mr Haworth, who stated that he possessed several English specimens It is closely allied to P Alexis The wings measure rather more than an inch and a quarter in expanse, and are of a less brilliant blue than in Dorylas, with a slender black border, which is extended slightly upwards along the veins The fringe is white, nearly as in Alexis The under surface of the four wings is of an ashy-grey, with the base powdered with bluish green, and with a transverse row of ocellated spots preceded in the hind wings by four or five fulvous lunules, which become more distinct as they approach the anal angle These lunules are angulated above with black, and marked below with black spots At the extremity of the fore wings, in the place of the fulvous, submarginal band, are the rudiments of several greyish lunules The triangular spot at the tip of the discoidal cell is arched, and not preceded by any other spot, whilst that of the hind wings is preceded by a row of three black spots The breast and feet are bluish, the thorax and upper side of the abdomen blue, and the antennæ ringed with white

The upper side of the female is of a shiny brown colour, with a marginal row of fulvous lunules, each resting upon a black spot this row of lunules occasionally extends more or less along the edge of the fore wings which moreover, are marked with a black spot at the extremity of the discoidal cell The under side is of a yellowish or reddish-grey, scarcely saturated with blue at the base, with the markings more distinct than in the male, and the marginal, fulvous band distinct upon the fore wings but not so strong in the fore as in the hind wings

At the time when Mr Stephens published his Illustrations, he did not possess this species and in his more recent manuscripts (with the sight of which he has obligingly favoured me) I find this species indicated with marks of doubt as being in his own collection The specimens in his collection, from which our figures 1—3 of plate 36 are derived, appear to me to be but varieties of P Alexis The fore wings of the male on the under side (fig 3) have two ocelli at the base, preceding the central spot, and the female is powdered with blue (fig 2,) and has the fulvous marks of the margin of the fore wings spotted with black, none of which characters agree with the true Icarius How far Mr Haworth's "several English specimens," which were taken in Kent, agreed with the true Icarius, I have now no means of ascertaining

On the Continent, Icarius is found in the Pyrenees, the Piedmontese Alps, Lapland [?], Hungary and some parts of Germany

SPECIES 13.—POLYOMMATUS AGESTIS THE BROWN ARGUS BUTTERFLY

Plate xxxvi fig 5 7

SYNONYMES —*Papilio Agestis*, Wien Verz, Hubner, Pap pl 62, f 303, 304
Polyommatus Agestis, Latreille, Stephens, Duncan, Brit Butt, pl 34, fig 1 Wood, Ind Ent t 3, f 9, and t 2 f 72
Agriades Agestis, Hubner, Verz bek Schm
Papilio Idas Lewin Pap, pl 39, f 1, 2 Donovan, Brit Ins, vol 10, pl 322, f 2 Haworth, Jermyn (not *Idas*, Linn, F S, which is the female of Argus)
Lycaena Idas, Ochsenheimer Leach
Pap Medon Esper, Pap pl 32, Suppl 8, f 1

We have now taken leave of the species of Polyommatus, in which the males are ornamented with blue or purple tints on the upper surface of the wings In both sexes of this species the wings are coloured alike being of a fine silken brown with a very slender pale margin along the costa, and with a row of small bright orange-coloured lunulated spots, marked on the outside in the hind wings with small black round dots There is also a small black crescent at the extremity of the discoidal cell of the fore wings The fringe is white or pale brown, with minute dark lines at the extremity of the veins The upper surface of the body is black, with greyish hairs The under side of the wings is of a brownish ash colour, the fore wings with a rather large and very distinct white spot generally inclosing a smaller black one at the extremity of the discoidal cell, succeeded by a *strongly curved* row of five or six (the inner one when present being minute and duplicated) similar ocelli These are succeeded by the same number of fulvous patches edged within with a brown curve, and marked next the margin with a brown spot The margin is slender, and blackish brown, dilated at the tips of the veins, and white within, and the fringe white, slightly marked with brown at the extremity of the veins The hind wings are tinted with blue at the base, which is marked with three ocelli, the discoidal white spot is transverse-oval, and emits a small branch behind, it is but slightly marked with black in the centre, beyond this is a curved and irregular row of eight white spots, varying in size (the middle ones being confluent), each of which has a black dot in the centre Then follows a row of slender, black, pointed arches, edged within with a slender white line, and externally bearing a row of fulvous patches succeeded by a row of white, connected, transverse-oval spots each bearing a small black dot, the margin is very slender, and black, and dilated at the extremity of the veins

The female differs in not being so intensely brown on the disk of the wings, and in having the fulvous band of spots larger and more distinct, extending to the front margin of the fore wings, and more strongly marked with black spots on the hind wings Varieties occur in the number of the spots on the under side of the wings, and in the size and extent of the row of fulvous spots

The expansion of the wings in this species varies from an inch to an inch and a quarter

This species appears to be double-brooded, May and June, and July and August, being the times of its appearance in the winged state It is found in most of the southern counties of England in tolerable abundance, as well as in various localities in Norfolk and the adjacent counties The most northerly recorded situation is Seaham Dean near Sunderland

The caterpillar is green, with a pale, angulated row of dorsal spots, and a central brownish line. It feeds, according to Haworth, on grasses, but Esper figures it upon the wild strawberry. It appears in this state in the months of April and June.

In the Entomological Magazine (vol ii p 515) Mr Newman endeavoured to prove that this species and the two following "are but one species. Specimens taken at Ramsgate, Dover, Hythe, Hastings, Rye, Brighton, Worthing, Little Hampton, Chichester, Portsmouth, Isle of Wight, Dorsetshire and Somersetshire exhibit the "typical form" of the species (as described above), at Birmingham, Worcester and Shrewsbury "an evident change has taken place, the band of rust-coloured spots has become less bright, at Manchester these spots have left the upper wing almost entirely, at Castle Eden Dean they are scarcely to be traced, and a black spot in the centre of the upper wing becomes fringed with white, in some specimens it is quite white, the butterfly then changes its name to Salmacis. We proceed further northward, and the black pupil leaves the eyes on the under side, until at Edinburgh they are quite gone, then it is called Artaxerxes. Mr Stephens does not, however, agree with Mr Newman in this respect, stating that "his definitions do not accord with my series of specimens of the three insects obtained from nearly every one of the localities enumerated by him" (Illustr Haust 4, p 382) Boisduval also gives the Artaxerxes as distinct, stating that its fore wings are proportionably longer than those of Agestis, to which I may add that the relative position of the spots (which seems to me in this genus to afford a good specific character) is different in the two species, especially on the under side of the upper wings.

DESCRIPTION OF PLATE XXXVII

INSECTS —Fig 1 Polyommatus Salmacis (the Durham Argus B) 2 The female 3 Showing the under side

" Fig 4 Polyommatus Artaxerxes (the Artaxerxes B) 5 The female 6 Showing the under side

PLANTS —Figs 7 and 8 Ononis procurrens (trailing Rest harrow)

The insects on this plate are all figured from the cabinet of Mr Stephens. I have placed these two species together to [illegible] the collector to compare them, as they have been supposed by some to be merely variations of one species. In Salmacis it will be seen the female is sometimes without the white mark, which is more constant in Artaxerxes, and Salmacis has more frequently a slight indication of an orange border on the fore wings. On the under side Artaxerxes is without the black pupil in the pale spots, which is pretty constant in Salmacis. H N H

SPECIES 14 —POLYOMMATUS SALMACIS THE DURHAM ARGUS BUTTERFLY

Plate xxxvii fig 1—3

SYNONYMES —*Polyommatus Salmacis*, Stephens, Illust Haust vol i p 235
Wood, Index Ent t 3, p 73, ♂ ♀, and fig 12 Duncan, Brit Butt pl 31, fig 2 and 3

This species is intermediate between Agestis and Artaxerxes and varies in the expanse of its wings from 1 [illegible] to 1⅛ inches. The upper side of the wings is of a silky blackish brown colour, with a black spot at the extremity of the discoidal cell in the males, and a white one in the females, which is, however, sometimes obsolete especially in the latter sex. There is also a row of submarginal fulvous spots in the hind wings, which sometimes also extends along the margin of the fore wings, but it is occasionally almost obsolete in the male. The fringe is white, with slight brown marks at the base. The under side of the wings is of a brownish grey, the anterior wings having a white spot at the extremity of the discoidal cell, succeeded by a curved row of similar spots, each marked in the centre with a dusky point, there is also a submarginal row of orange spots bounded above with a dusky crescent, and marked beyond with a dusky spot surrounded with white, the extreme margin being marked by a dusky line the posterior wings have several white spots towards the base, a larger discoidal one, a curved irregular row of white tuberculated spots beyond the middle, with a broad patch of white connecting the middle spots with the submarginal band of fulvous spots, which are similar in their markings to the corresponding row of the fore wings the white subocellated spots on this side of the wings have the middle marked more conspicuously with a dusky spot in the females than in the males

Mr Stephens short description of this species is as follows —"Alis fusco-nigris, subtus fuscescentibus maculis subocellatis, anticis supra in masculis puncto discoidali atro in fœminis albo, posticis utrinque fascia submarginali rubra," and Mr Wailes, an acute entomologist, resident upon the spot where the species occurs, namely Castle Eden Dean, near Durham, (it also occurs in the magnesian limestone district, near Newcastle,) entirely coincides with Mr Stephens in considering it as a distinct species, observing, however, that Mr S's description is not quite correct since out of at least 150 specimens, the variety with the black spot forms two-thirds of the whole, and that neither sex possesses exclusively either the white or black spot, though the majority of the former variety are males (Entomol Mag i p 42) He further states that this butterfly appears to be confined to the sea-banks, having only seen a few stragglers so far from the coast as half a mile It appears in July

---

SPECIES 15 —POLYOMMATUS ARTAXERXES THE SCOTCH ARGUS

Plate xxxvii fig 4—6

SYNONYMES —*Hesperia Artaxerxes*, Fabricius, Lewin, Pap pl 39 fig 8—9 Haworth Donovan Brit Ins, v 16, pl 541 (Papilio Art)

Lycæna Artaxerxes, Leach

Argus Artaxerxes Boisduval, Icon Hist Lep pl 11 fig 7—8

Polyommatus Artaxerxes, Stephens Jermyn, Wood Ind Ent pl v f 71 and 13 Duncan Brit Butt pl 31, fig 4

This species is closely allied to the preceding, if, indeed, it be not a local variety of it It varies in the expanse of the wings from an inch to an inch and a sixth The upper surface of the wings in both sexes is of a silky-blackish brown colour, with the anterior margin very slender and white, and a small white spot at the extremity of the discoidal cell, and in a few instances, a similar minute white spot occurs on the disk of the hind wings above There is also a submarginal row of small orange-red marks on all the wings, although they are often almost, or even entirely obsolete in the fore wings The fringe is white, slightly marked with brown at the base opposite to the extremity of the veins The under side of the wings is of a greyish brown, the anterior wings having a large round patch at the extremity of the discoidal cell, succeeded by a *slightly curved* row of white oval spots beyond this is a row of fulvous spots, bounded within by dusky crescents edged with white, and terminating in round black spots, beyond which is a slender bar of white immediately preceding the slender dark margin which is dilated at the extremity of the veins The under side of the fore wings is marked by four white spots, forming an oblique line at the base, the spot at the extremity of the discoidal cell is white and of transverse-oval form, emitting a minute straight branch from its outer edge, near this, on the costa of the hind wings, are two white spots and there are six others, forming an interrupted white bar beyond the middle of the wings those in the centre being confluent, and thus forming a larger white patch, the margin of this pair of wings is ornamented with fulvous spots and other markings exactly corresponding with those on the margin of the fore wings

The female, as in Agestis, closely resembles the male, but the submarginal row of fulvous spots is generally larger and more extended into the fore wings than in the male The upper side of the antennæ is like that of Agestis, except that the extremity of the club is reddish, their under side is almost entirely whitish

No. XIV.] [Price 3s.

BRITISH INSECTS AND THEIR TRANSFORMATIONS.

BRITISH BUTTERFLIES

AND

Their Transformations.

ARRANGED AND ILLUSTRATED IN A SERIES OF PLATES

BY H. N. HUMPHREYS, ESQ.

WITH CHARACTERS AND DESCRIPTIONS

BY J. O. WESTWOOD, ESQ., F.L.S.,

SEC. OF THE ENTOMOLOGICAL SOCIETY, ETC. ETC.

LONDON:
WILLIAM SMITH, 113, FLEET STREET.

MDCCCXLI.

BRADBURY AND EVANS, PRINTERS, WHITEFRIARS.

WORKS PUBLISHED BY WILLIAM SMITH, 113, FLEET STREET.

SMITH'S STANDARD LIBRARY.

In medium 8vo, uniform with Byron's Works, &c

Works already Published

ROKEBY. BY SIR WALTER SCOTT 1s 2d.

THE LADY OF THE LAKE 1s
THE LAY OF THE LAST MINSTREL 1s
MARMION 1s 2d
THE VISION OF DON RODERICK, BALLADS AND LYRICAL PIECES By SIR WALTER SCOTT 1s
THE BOROUGH By the REV G CRABBE 1s 4d.
THE MUTINY OF THE BOUNTY 1s 4d
THE POETICAL WORKS OF H KIRKE WHITE 1s
THE POETICAL WORKS OF ROBERT BURNS 2s 6d
PAUL AND VIRGINIA, THE INDIAN COTTAGE, and ELIZABETH 1s 6d
MEMOIRS OF THE LIFE OF COLONEL HUTCHINSON, GOVERNOR OF NOTTINGHAM CASTLE during the Civil War By his Widow, Mrs LUCY HUTCHINSON 2s 6d
THOMSON'S SEASONS, AND CASTLE OF INDOLENCE 1s
LOCKE ON THE REASONABLENESS OF CHRISTIANITY 1s
GOLDSMITH'S POEMS AND PLAYS 1s 6d
——— VICAR OF WAKEFIELD 1s.
——— CITIZEN OF THE WORLD 2s 6d
——— ESSAYS, &c 2s
*** These four Numbers form the Miscellaneous Works of Oliver Goldsmith, and may be had bound together in One Volume, cloth-lettered, price 8s
KNICKERBOCKER'S HISTORY OF NEW YORK 2s 3d
NATURE AND ART. By MRS INCHBALD 10d
A SIMPLE STORY By MRS INCHBALD 2s
ANSON'S VOYAGE ROUND THE WORLD 2s 6d
THE LIFE OF BENVENUTO CELLINI 3s
SCHILLER'S TRAGEDIES THE PICCOLOMINI, and THE DEATH OF WALLENSTEIN 1s 8d
THE POETICAL WORKS OF GRAY & COLLINS 10d
HOME By MISS SEDGWICK 9d
THE LINWOODS By MISS SEDGWICK 2s 8d
THE LIFE AND OPINIONS OF TRISTRAM SHANDY By LAURENCE STERNE 3s
INCIDENTS OF TRAVEL IN EGYPT, ARABIA PETRÆA, AND THE HOLY LAND By J L STEPHENS, ESQ 2s 6d
INCIDENTS OF TRAVEL IN THE RUSSIAN AND TURKISH EMPIRES By J L STEPHENS, Esq 2s 6d
BEATTIE'S POETICAL WORKS, and BLAIR'S GRAVE 1s
THE LIFE AND ADVENTURES OF PETER WILKINS, A CORNISH MAN 2s 4d
UNDINE A MINIATURE ROMANCE Translated from the German, by the REV THOMAS TRACY 9d
THE ILIAD OF HOMER Translated by ALEX POPE 3s
ROBIN HOOD, a Collection of all the ancient Poems, Songs, and Ballads now extant, relative to that celebrated English Outlaw, to which are prefixed, Historical Anecdotes of his Life Carefully revised, from RITSON 2s 6d
THE LIVES OF DONNE, WOTTON, HOOKER, HERBERT, AND SANDERSON Written by IZAAK WALTON 2s 6d
THE LIFE OF PETRARCH By MRS DOBSON 3s
MILTON'S PARADISE LOST 1s 10d
——— PARADISE REGAINED, and MISCELLANEOUS POEMS 2s
RASSELAS By DR JOHNSON 9d
ESSAYS ON TASTE By the Rev ARCHIBALD ALISON, LL B 2s 6d
THE POETICAL WORKS OF JOHN KEATS 2s
THOMSON'S POEMS AND PLAYS, COMPLETE 3s
PICCIOLA A Tale from the French 1s 4d
THE POETICAL WORKS OF DR YOUNG, COMPLETE 5s
VOLTAIRE'S HISTORY OF CHARLES XII OF SWEDEN 1s 10d
THE POETICAL WORKS OF WILLIAM CULLEN BRYANT 1s

FAUST from the German of GOETHE A new Translation, by LEWIS FILMORE, Esq 1s 6d

FOUR VOLUMES ARE NOW ARRANGED ON THE FOLLOWING PLAN —

ONE VOLUME OF "POETRY,"

CONTAINING

SCOTT'S LAY OF THE LAST MINSTREL
SCOTT'S LADY OF THE LAKE
SCOTT'S MARMION
CRABBE'S BOROUGH
THOMSON'S POETICAL WORKS
KIRKE WHITE'S POETICAL WORKS
BURNS'S POETICAL WORKS.

A SECOND VOLUME OF "POETRY,"

CONTAINING

MILTON'S POETICAL WORKS
BEATTIE'S POETICAL WORKS
BLAIR'S POETICAL WORKS
GRAY'S POETICAL WORKS
COLLINS'S POETICAL WORKS
KEATS'S POETICAL WORKS
GOLDSMITH'S POETICAL WORKS

ONE VOLUME OF "FICTION,"

CONTAINING

NATURE AND ART By MRS INCHBALD
HOME By MISS SEDGWICK
KNICKERBOCKER'S HISTORY OF NEW YORK By WASHINGTON IRVING
PAUL AND VIRGINIA By ST PIERRE
THE INDIAN COTTAGE By ST PIERRE
ELIZABETH By MADAME COTTIN
THE VICAR OF WAKEFIELD By OLIVER GOLDSMITH.
TRISTRAM SHANDY By LAURENCE STERNE.

A SECOND VOLUME OF "FICTION,"

CONTAINING

A SIMPLE STORY By MRS INCHBALD
PICCIOLA FROM THE FRENCH OF X B SAINTINE
THE LINWOODS By MISS SEDGWICK
THE LIFE AND ADVENTURES OF PETER WILKINS
UNDINE FROM THE GERMAN OF FOUQUE

ONE VOLUME OF "VOYAGES AND TRAVELS,"

CONTAINING

ANSON'S VOYAGE ROUND THE WORLD
BLIGH'S MUTINY OF THE BOUNTY
STEPHENS'S TRAVELS IN EGYPT, ARABIA PETRÆA, AND THE HOLY LAND
STEPHENS'S TRAVELS IN GREECE, TURKEY, RUSSIA, AND POLAND

These Volumes may be had separately, very neatly bound in cloth, with the contents lettered on the back, price 10s 6d each

Those parties who have taken in the different works as they were published, and who wish to bind them according to the above arrangement, may be supplied with the title-pages gratis, through their booksellers, and with the cloth covers at 1s each

Like several of the allied species, this insect is double-brooded, appearing in June and August. Fabricius, by whom the species was first described, gave it as a native of *England*, from the drawings of Mr. Jones. It appears, however, to be exclusively an inhabitant of Scotland, and was, until lately, supposed to be only found on Arthur's Seat, Edinburgh, being of such rarity, that scarcely a single cabinet possessed a specimen, the owners endeavouring, according to the bad taste of the day, to supply its place with a drawing of the insect stuck in their cabinets. More recently, however, it has occurred in considerable plenty, not only in the locality above mentioned, but also on Salisbury Crags, King's Park, and near Duddingston Loch, Pentland Hills, near Queensferry and Rosslyn Castle, Jardine Hall, Dumfriesshire, and Flisk, in Fifeshire. The other localities mentioned by Mr. Stephens appear to belong to Salmacis, except the last, "Dartmoor, 23 August, 1823, Dr. Leach," which is probably erroneous.

In addition to what has been already observed under P. Agestis, relative to the specific rank of this and the two preceding species, it must be stated, that although Agestis is very abundant on the Continent, the Continental Entomologists have never met with a single specimen of Artaxerxes, their cabinets being entirely furnished with Scotch specimens.

Polyommatus Titus (Hesperia T., Fabricius, Turton, Donovan in Rees' Encycl. art. Papilio, Jermyn, Stephens,) was described by the first-named author as a native of England on the authority of Mr. Drury and Mr. Jones's collection of drawings of Lepidoptera.* Mr. Haworth, however, who was personally well acquainted with the latter gentleman, states that the information given to Fabricius was incorrect, the insect not being a native of this country. The following is a translation of his character:—Stature quite like Artaxerxes, &c.; all the wings above, brown, unspotted; beneath, also brown; the anterior, with a posterior or submarginal, row of short white and black lines, and the posterior, with a central short line, and a row of black dots, edged with white; near the margin is a row of red spots, each marked with a black dot.

FAMILY V.

HESPERIIDÆ.

This family, which is the sixth in the arrangement of the diurnal Lepidoptera in general, was well indicated by Linnæus, under the title of Papiliones Plebeii Urbicolæ, and is composed of a very distinct tribe of butterflies, constituting, indeed, a primary division amongst them, which Boisduval has termed Involuti, from the circumstance of the caterpillars inclosing themselves in a curled-up leaf; and thus, as well as in several other important characters, approaching the moths.

The six feet are of uniform size in both sexes; the hind tibiæ have a pair of spurs at the apex, and generally another pair near the middle of the limb, a character found in none of the preceding butterflies; the hind wings are generally horizontal during repose, and in some species all the wings are placed in this manner (Tamyris Zeleucus,

* This collection of drawings is still in existence, being in the possession of a gentleman resident at Chichester; many of the species figured in it, and thence described by Fabricius, have since been published by Donovan in his Naturalist's Repository.

Fab Swainson Zool Ill, vol I, pl 33) The antennæ are wide apart at the base, and are often terminated in a very strong hook, the labial palpi have the terminal joint very small, the spiral tongue (or maxillæ) is very long, and the discoidal cell of the hind wings is not closed

The caterpillars, of which, however, but few are known, are cylindrical, without spines, with the anterior segments narrowed, and the head very large, which thus appears to be borne upon a foot-stalk, the hind part is always obtuse These larvæ* roll up leaves, in which they construct a very slender silken cocoon, wherein they are transformed to chrysalides, which are entire and without angular prominences

Poey in his Centurie des Lépidoptères de Cuba, pl 1, and Swainson in his Zoological Illustrations, vol I, Abbot and Smith, Reaumur, Stoll, Hubner, &c, have represented the transformations of various species The chrysalis is generally smooth, but occasionally angulated, of a lengthened form, and attached at one end as well as girt round the middle, the transformation being effected in the rolled-up leaf which served the caterpillar as its abode

The species are of comparatively small size, and of obscure colours, but many are ornamented with pellucid spots, and others have the hind wings furnished with long tails The body is short, very robust, and their flight is accordingly very strong, rapid, and so peculiar that they have obtained the name of skippers,—indicative of their singular short, jerking kind of flight They also frequently settle on flowers, leaves, or branches, as well as upon the ground, with which their dull colours well associate

H Tages (according to Dr Abbot, Linn Trans vol 5, p 276) flies early in the morning, its flight being extremely short, and very near the ground Mr Curtis mentions the curious circumstance, that old specimens, whilst alive, frequently lose one or both of their palpi, an accident he had only observed amongst the Pyralidæ

The relations of these butterflies with other insects are very interesting Latreille united in the same group a singular exotic genus, Urania, which is nearly related to the Hesperi-sphingides Mr Swainson, indeed, states that their *palpable affinity* to the *hawk-moths* has induced almost every writer to place them as the connecting link between the diurnal and crepuscular Lepidoptera, but such is not the case, it is with the Hesperi-sphingides that they are nearest allied, their relationship with the hawk-moths being very slight They have another relation with the Tortricidæ, a family of small moths, founded merely upon the habit which is exhibited by the caterpillars of both groups of rolling up the leaves of plants, and which, in the Tortricidæ, becomes a practical source of annoyance in many instances These relations are self-evident, but another has recently been pointed out, which appears to me to be so far-fetched and ridiculous as to merit only silence, were it not that it forms part of a system which is asserted to be all natural, and which must supersede all others hitherto or hereafter to be promulgated. "The Natural Arrangement of Insects" of Mr Swainson furnishes us (p 205) with a test of the relative position in nature of the tribes of the Coleoptera, founded upon the corresponding position of the families of the diurnal Lepidoptera. The Hesperidæ are thus made to represent the Malacoderm-beetles, because the skin of the larvæ of the former is so thin that the caterpillars are obliged to defend themselves by the artificial

---

* "The larvæ of the Hesperidæ are so strikingly distinguished from those of the Polyommatidæ, and the only one known of the Erycinidæ, that it is really surprising how entomologists still continue to arrange them in the same group"—Swainson, Hist of Insects, p 97 Nothing more completely proves the ignorance of this writer of the works of modern entomological writers, by the majority of whom Hesperia is constituted into a separate family

means of curled-up leaves "The Hesperidæ are, in fact, the *soft-skinned butterflies*, just as the Malacoderms among the Coleoptera" * ‡

England is comparatively poor in the species of this family, tropical America being the metropolis of the group, at least 300 species having been collected in that part of the world Other parts of the world, as India, New Holland, South Africa, and Europe, possess various species, but fewer in number Boisduval thinks that there are more than 400 species in collections

The species are very closely allied together, and difficult to be determined, except by very precise examination M Rambur has, however, proved in the last number of his Faune Entomologique de l'Andalousie just published, that good specific distinctions exist between nearly-allied species in the structure of the male organs of generation

The study of the whole of this extensive family can alone determine the propriety of the distribution of the species into genera or still minor groups It is impossible to examine the very few indigenous species we possess without being convinced of the difficulty of attempting this from so small a portion of the group For instance, the antennæ in Malvæ or Alveolus, and Tages, have the club differently formed, and the position of their wings in repose is different, although they agree in the folded costa of the fore wings of the males, and in the curved clava of the antennæ Again, the club of the antennæ differs in its form in every one of the species composing the genus Pamphila of Stephens, and yet this is the character which the last-named author uses to characterise the two genera into which he has divided the British species Hubner, Boisduval, and still more recently Zeller, have attempted the distribution of the species into subordinate groups, and it is much to be regretted that Mr Swainson's researches in this difficult family have not yet been published

DESCRIPTION OF PLATE XXXVIII

INSECTS —Fig 1 Pyrgus Malvæ (the Grizzled Skipper) 2 The female 3 Showing the under side 4 The Caterpillar 5 The Caterpillar prepared for its change to the chrysalis state 6 The chrysalis

,, Fig 7 The white banded variety of Pyrgus Malvæ, by some considered a distinct species 8 Showing the under side

,, Fig 9 Nisoniades Tages (the Dingy Skipper) 10 The female 11 Showing the under side 12 The Caterpillar 13 The Chrysalis

,, Fig 14 Pyrgus Oileus 15 Showing the under side

PLANTS —Fig 16 Dipsacus fullonum (Fullers' Teazle)

,, Fig 17 Eryngium campestre (Field Eryngo)

P Malvæ and the variety are from specimens in the British Museum N Tages from specimens in the cabinet of E Doubleday, Esq of Epping, and P Oileus from specimens taken by Mr Doubleday in North America he has no doubt that it is the insect which has been considered a British species, and found its way into some British cabinets under the name of P Oileus The caterpillar and chrysalis of P Malvæ are from Hubner and those of N Tages from Godart —H N H

PYRGUS, HUBNER (THYMELE P STEPHENS)

The species of this genus, or rather sub-genus, are distinguished by the greater length of the palpi, which are very hairy, and extend in front of the head, being at least as long as the head, the terminal joint being slender,

* In like manner the Erycinidæ represent the Malacoderm beetles, because the larva of the former resembles that of a Cassida or Tortoise beetle, and the Satyridæ [Hipparchides] represent the Capricorn beetles, because the antennæ in the perfect insects of both groups are moderately long, and because the head of the larva of these butterflies is often armed with long horns!! Scientific trifling can scarcely go further than this

distinct, and exserted. The antennæ are rather short, without any hook at the tip, and terminated by a gradually-formed arched club; the head is rather broad, with a tuft of recurved hair at the base of the antennæ, and the thorax robust. The wings are short, and rounded along the outer margin, in both sexes; the front margin towards the base in the males being folded, the base rounded; the mediastinal vein scarcely extends beyond the middle of the front margin of the wing; the postcostal one extends to a short distance below the apex of the wing, emitting on its front side four straight branches, the fourth of which runs to the tip of the wing; it also emits a branch from its posterior side. The great median vein is divided into three branches, and between the anterior one of these and the posterior one of the postcostal vein, is a straight free vein.* The males are not distinguished by having a thickened, oblique patch upon the disc of the fore wings. The wings in repose are *deflexed*. The abdomen in the males is narrow, with the tip bearded, whilst in the females it is more robust, with the tip acute and nearly naked. The cilia of the wings is long, alternately black and white, and the wings are also of a dark colour, spotted with white.

The larvæ are naked, or but very slightly pubescent, resembling those of the Tortricidæ, with a large head, the following segment being attenuated; generally subsisting upon the rolled up leaves of malvaceous plants. The pupa is entire and conical in its form, inclosed in a cocoon, and fastened by the tail as well as by a girth round the middle.

As there are a considerable number of species agreeing with Malvæ, I have retained that as the type of a distinct genus, for which I have employed Hubner's name in preference to that of Thymele of Fabricius (used by Stephens), the real types of which are exotic-tailed species, and because it has a priority of date over that of Syrichtus of Boisduval, employed for the group.

SPECIES I.—PYRGUS MALVÆ. THE GRIZZLED SKIPPER.

Plate xxxviii. fig. 1—6.

Papilio Malvæ, Linnæus, Faun. Suec. 1081; Lewin, pl. 46, f. 8—9; Haworth; Turton; Harris, Aurelian, pl. 32, fig. i—m.
Hesperia Malvæ, Leach; Curtis; Dalman, Hesp. Su. 202. 6; Zetterstedt, Faun. Lapp. p. 915 (not Thymele Malvæ of Stephens and Wood, nor of Fabricius).
Syrichtus Malvæ, Boisduval, Icon., p. 231.
Papilio Alveolus, Hubner, Pap. t. 92, f. 466—467.
Thymele Alveolus, Stephens; Duncan Brit. Butt. v. 2, pl. 1, f. 1; Wood, Ind. Ent. t. 3, f. 75.
Pyrgus Alveolus, Hubner (Verz. bek. Schmett.).
Papilio Sao, Bergstrasser, Fur. Schmett. t. 10, f. 8; Faun. Franc. pl. 26, f. 7, 8.
Hesperia Fritillum minor, Fab. Ent. Syst. 3, part 1, p. 351, pl. 336.
Papilio Fritillum, Lewin, pl. 46, f. 4, 5 (variety).
Papilio Lavateræ, Fabricius, Haworth, Jermyn (variety). Not P. Lavateræ of Hubner.
Papilio Altheæ, Borkhausen (variety).
Papilio Malvæ minor, Esper.

This species generally measures about an inch in the expanse of the wings, varying a little both more or less. On the upper side the wings are of a dark brown colour, marked with many small, squarish, cream-coloured spots, of which there are about fourteen on each of the fore wings, the ground colour of which, especially towards the base, is much powdered with white, especially in the males. The middle of the hind wings is marked with several more or less confluent larger spots, and beyond the centre is a curved row of six small dots. The cilia is white, spotted alternately with black; the body has a greenish hue. Beneath, the ground colour of the wings is much paler, the spots towards the tips of the wings forming fine lines; the spots are also larger and more

---

* I have found no material variation in regard to the arrangement of the veins of the wings, in any of the indigenous species of the entire family.

numerous on the hind wings, especially towards the base, and the front margin has a large white blotch, the veins in these wings are also pale coloured

A not very uncommon variety, regarded by Fabricius, Lewin, &c as distinct, is represented in our figures 7 and 8, in which there is a white oblong blotch in the middle of the fore wings towards the posterior margin visible on both sides, which is frequently duplicated from the confluence of two contiguous spots The white dots are also longer and larger than in the typical individuals Mr Stephens possesses a specimen with one of the fore wings marked as in the variety, and the other as in the type

The caterpillar is green, with pale longitudinal stripes, a black head, and a yellow ring round the neck It feeds on the teazle, the leaves of which it rolls up

This is a common species, occurring in woods and dry pastures in Kent Surrey, Essex, Hertford, Wilts, Durham, Cambridge, Northumberland, and the south of Scotland It appears at the end of May Reaumur has given the history of this species in the eleventh plate of his first volume The Rev W T Bree informs us, that he once took the "variety?" regarded by some writers as distinct under the name of Lavaterae, near Yarmouth, in the Isle of Wight, and that a friend takes it in some abundance in the Forest, near Bewdley Worcestershire "It seems to be, like the white Colias Edusa, what may be called a permanent variety, or one which is constantly occurring"

I have followed Boisduval and Zetterstedt in restoring to this species the name of Malvæ, that name in the hand-writing of Linnæus himself being attached to his specimen of this insect in the Linnæan Cabinet His words also, "margine *quasi dentato*, interjacentibus maculis albis," distinguish it at once from the following species

---

SPECIES 2—PYRGUS OILEUS?

Plate xxxviii fig 13—14

These figures represent a North American insect, respecting the history of which, as a doubtful inhabitant of this country, it will be necessary to give the following details

The late Mr Haworth, in the 3rd part of the Entomological Transactions (p 334), gives the following statement —

"Oileus Papilio (The *Georgian Grizzle*), Gmel, Syst Nat 2370, 269?

"Obs —Has been caught in Bedfordshire by the Rev Dr Abbott and is in Leman's ancient English Cabinet, now in the possession of Lee Phillips, Esq, Manchester"

From this English name, it is evident that Mr Haworth considered the specimens as identical with a North American species, to which he applied Gmelin's name Oileus, but with a mark of doubt, which is by no means surprising, when it is stated that Gmelin gives Algiers as the locality of that insect with only the following short description of it —"P alis subdenticulatis fuscis albo maculatis, supra basin exteriorem primorum linea alba Pithoni simillimus (Syst Nat 4, p 2370)

Mr Stephens describes the species as having the "wings rounded, anterior varied with black and white, posterior beneath cinereous with waved black streaks antennæ black, the club cinereous beneath" and suggests that the specimens in question (which he had not seen) may be rather identical with the P Fritillum of Hubner Mr Curtis however, states, that they all agree with the North American species

DESCRIPTION OF PLATE XXXIX

INSECTS —Fig 1 Pyrgus? Malvarum 2 Showing the under side 3 The Caterpillar when young 4 When in a more advanced stage 5 The Chrysalis

" Fig 6 Cyclopides Paniscus 7 The female 8 Showing the under side 9 The Caterpillar

" Fig 10 Cyclopides Sylvius 11 The female 12 The under side of the male

PLANTS —Fig 13 Althæa officinalis (the marsh mallow)

" Fig 14 Plantago major (the greater plantain)

Pyrgus Malvarum is from a specimen which I took in Italy, near Turin, Mr Stephens does not consider it has any claim to be considered an English species but as it is found in old collections I have thought it right to give a figure of it The Caterpillars are from Godart, who minutely describes rearing them himself Cyclopides Sylvius has no greater claim to be considered British than P ? Malvarum, but I have introduced it for similar reasons It is probably a Continental variety of C Paniscus, and it is not impossible that some English varieties may have been taken nearly resembling it C Paniscus is from specimens kindly furnished by Mr Doubleday, and the Caterpillar is from Godart H N H

SPECIES 3 —PYRGUS ? MALVARUM

Plate xxxix fig 1—5

SYNONYMES —*Hesperia Malvarum* Hoffmansegg, Ochsenheimer
Hesperia Malvæ, Fabricius, Stewart Donovan, 16, pl 567
Thymele Malvæ, Stephens, Wood, Ind Ent t 53, f 17
Carcharodus Malvæ, Hubner, Verz bek Schmett (but not P Malva Linn)
Papilio Althææ, Hubner (variety)
Hesperia de la Guimauve, Godart

The Linnæan name of Malvæ having been applied to this species, and given as a native insect by Stewart and Donovan, it has become necessary to introduce it into this work to show the distinction between the two species, in fact, the present is more nearly allied to Hesperia Tages, its dentated wings, however, induced Hubner to place it in a separate group under the name of Carcharodus The wings are *dentated* and brown with waving cinereous lines, and six transparent spots on the fore wings, the hind wings beneath dotted with white Of this species no authentic instance has occurred of the capture in this country The caterpillar is of a dirty blackish green, when young, but becomes lighter coloured as it increases in age, with darker longitudinal lines, a black head, and a yellow neck It feeds on mallows

---

NISONIADES, HUBNER (THANAOS, BOISDUVAL)

This group differs from the preceding in having longer, slenderer antennæ, with the club attenuated at the tip, the palpi with the last joint thicker, the anterior margin of the fore wings slightly angulated beyond the middle, the surface of the wings not tessellated, and the fringe alternately spotted. It agrees with it in having the wings identical in the outline in both sexes, and in the costa being folded at the base in the males The wings are deflexed in repose, and they are longer than the abdomen The difference in their colour in the two sexes indicates a relation with the following group As there are several species which agree in these respects, I have adopted the group with Hubner's name, which is prior to that of Boisduval

SPECIES 1.—NISONIADES TAGES. THE DINGY SKIPPER.

Plate xxxviii. fig. 9—13.

SYNONYMES.—*Papilio Tages*, Linnæus. Lewin, Pap. pl. 45, f. 3, 4. Haworth, Harris Aurelian, pl. 34, fig. 0.
Hesperia Tages, Fabricius, Leach, Jermyn.
Thymele Tages, Fabricius (Gloss.), Stephens. Duncan Brit. B. t. 2, pl. 1, f. 2. Wood Ind. Ent. pl. 3, f. 76.
Thanaos Tages, Boisduval, H. n. Lep. pl. 9, B fig. 8.
Nisoniades Tages, Hubner (Verz. bek. Schm.)

The expansion of the wings of this species is about an inch and a quarter. The upper surface of the wings is brown, the fore wings marked with alternate waved bands of darker brown and grey, which in some specimens are in bright relief against each other, and separated by paler zigzag marks, in addition to which there are several indistinct whitish dots, one brighter than the rest being placed near the extremity of the costa; and there is also a marginal row of dull white dots. The hind wings are brown, with a small discoidal spot, beyond which are two rows of nearly obsolete paler dots. Beneath, the colour is uniformly greyish brown, the fore wings not shaded as above, but marked as well as the hind wings with the traces of the pale dots of the upper side. The male is duller and more uniformly coloured than the female.

The caterpillar is bright green, with the head brown, with yellow dorsal and lateral stripes dotted with black. It feeds on the field Eryngo, and bird's-foot lotus.

This species frequents woody pastures, heaths, &c., and is found at the beginning of May, in June, and the middle of July. It is by no means so common as Malvæ. It appears to be very widely extended, for in addition to the numerous localities in various parts of England given by Mr. Stephens, Mr. Duncan mentions several places in Scotland where it has also occurred.

CYCLOPIDES, HUBNER. (STEROPES, BOISDUVAL.)

The typical species of this group differs from the preceding in its long acuminated fore wings, short hind wings, scarcely bent club of the antennæ, and want of a fold at the base of the costa of the fore wings; and from the following, by the more slender body, differently formed club of the antennæ, and especially by the want of an oblique black patch across the middle of the wings of the males, and the identity of colouring in the sexes. The palpi are exserted, and as long as the head, with the terminal joint nearly concealed by hairs. The body in the males is long and slender, and slightly tufted at the tip. The antennæ are short, with the club stout, nearly straight, and not hooked at the tip. The wings are brown, tessellated with bright orange spots of a square or roundish form. A more important character, however, than any of the preceding, consists in the posterior tibiæ possessing only a pair of spurs at the tip.

Boisduval's name for this group is inappropriate, not only because there is a coleopterous genus Steropus, but also because one of the European Hesperidæ is also named Steropes.

SPECIES 1.—CYCLOPIDES PANISCUS. THE CHEQUERED SKIPPER.

Plate xxxix. fig. 6—9.

SYNONYMES.—*Hesperia Paniscus*, Fabricius, Ochsenheimer, Leach, Jermyn, Curtis.
Papilio Paniscus, Donovan, vol. 8, pl. 254, fig. 1. Haworth.
Pamphila Paniscus, Fabricius, Syst. Gloss., Stephens, Wood, Ind. Ent. t. 3, f. 77. Duncan, Brit. Butt. 2, pl. 1, fig. 3.
Steropes Paniscus, Boisduval, Hist. Lep. pl. 9 B. fig. 7.
Cyclopides Paniscus, Hubner (Verz. bek. Schmett.)
Papilio Brontes, Hubner.
Papilio Sylvius, Villars.

This pretty species is generally about an inch and a quarter in the expansion of its wings, which on the upper side are of a dark brown colour, spotted with orange; the exterior with a large orange blotch in the middle, marked towards the costa with a small square brown spot; beyond the middle is an irregular bar of orange, divided by the veins of the wings, and interrupted in the middle; the two small spots which are wanting to complete the bar being pushed outwards nearly to the margin of the wing, which is also marked with a row of fulvous dots. The hind wings are marked in the middle of the disc with three large round spots, and a submarginal row of smaller dots; the fringe is brown, the extremity being dirty orange. Beneath, the ground colour of the wings is tawny; the anterior with three discoidal and four smaller posterior dusky spots, which is also the colour of the veins at the extremity of the fore wings, and the entire veins in the hind wings, which are ornamented with pale buff spots, edged with brown; five being on the disc of larger size, and a submarginal row of smaller ones, the outer two of which are the largest. The antennæ on the under side are bright orange. The spottings differ in size in different specimens, but there is no material difference between the sexes.

The caterpillar has the head black, the neck with an orange ring; it is dark brown on the back, with two pale yellow stripes on the sides. It feeds on the Plantago major and Cynosurus cristatus. The perfect insect appears at the end of May. It is a very local species, although where found it is abundant. Castor Haglands Wood, near Peterborough; Clapham Park Wood, Bedfordshire; Whitewood, Gamlingay, Camb.; near Dartmoor; near Luton, Bedfordshire; a wood near Milton, Northamptonshire, are recorded by Curtis and Stephens as its localities; and the Rev. W. T. Bree informs us, that he took it abundantly the latter end of May 1823, in Barnwell Wolde, near Oundle, and in Rockingham Forest, and that he has also taken it near Woodstock. "In profusion in Monk's Wood, Hants, and in a wood near Oundle, Northamptonshire." H. Doubleday, Esq., in "The Entomologist," August 1841.

SPECIES 2.—CYCLOPIDES SYLVIUS.

Plate xxxix. fig. 10—12.

SYNONYMES.—*Papilio Sylvius*, Knoch, Hubner, Pap. pl. 94, f. 477, 478. Fuest, 1, pl. 71, Suppl. 20. f. 96. c. f.
Hesperia Sylvius, Fabricius, Ochsenheimer.
Pamphila Sylvius, Stephens. Wood, Ind. Ent. t. 3, f. 18.
Cyclopides Sylvius, Hubner.

This reputed British insect is nearly an inch and a quarter in the expanse of its fore wings, which are tawny orange above, spotted with black, four being on the disc, and a row of smaller ones along the margin, which is dusky. The hind wings, on the contrary, are brown above, spotted with orange, four spots being on the disc, and a row of five within the hind margin. On the under side the wings are nearly coloured as above, except that there is a chain-like series of brown spots, united by a black line on each vein with the outer margin of the fore wings. The hind wings have a similar submarginal series, the discoidal spots being the same as above.

A specimen of this species in Mr. Stephens's cabinet, obtained from "an old cabinet," in which it was named Paniscus, and "other specimens" in the late Mr. Milne's cabinet, which he also believed to be Paniscus, are the only authorities for the introduction of this species into our English lists; and it is more than probable that these were Continental specimens, introduced under mistake for the then rare Paniscus, with which their possessors thought them identical; indeed, Mr. Curtis entirely omits the species.

PAMPHILA, Fabricius. HESPERIA, Boisduval.

These insects are at once distinguished from all the preceding by the males possessing an oblique velvety patch of scales on the disk of the fore wings; moreover when in repose the fore wings alone are elevated. There is also a diversity in the colouring of the sexes, the females being brighter than the males. The head is large, as broad as the thorax, the eyes large and prominent, the palpi short, wide apart, very hairy, the last joint short, nearly naked, and exposed. The antennæ are terminated by a thick, nearly straight club, which is often furnished at the tip with a hook. The thorax is very robust, and the body as long as the hind wings. The wings are entire, with the fringe not alternated in its colours; the anterior ones are elongated, and the latter slightly sinuated at the anal angle, forming a short rudimental tail. A character which I have not seen noticed exists in the typical species, the outer margin of the fore wings of the females being much more rounded than in the males. The general colour of the wings is either tawny orange, marked with brown, or brown, strongly marked with the former colour; and generally the colours are so disposed as to leave a series of squarish spots near the outer margin of the fore wings.

The powerful flight of these insects far surpasses that of the other Hesperidæ, owing to the strength of their muscles and superior robustness of the body. In the larva state they generally feed upon low plants, especially Gramineæ. The exotic species are very numerous. The five species described in the following pages constitute several sections.

A. (Augiades, Hubn.) Antennæ hooked at the tip, head very large, palpi very short, squamose, last joint exposed; hind wings subtriangular. Vitellius? and its supposed female Bucephalus.

B. Head moderately large, palpi longer and hairy, hind wings more rounded, wings maculated, antennæ hooked at the tip. Comma and Sylvanus.

C. (Thymelinus, Hubn.) Head moderate, palpi moderately long and hairy, antennæ with the tip nearly straight, and not hooked at the tip. Linea and Comma.

DESCRIPTION OF PLATE XL.

INSECTS.—Fig. 1. Pamphila Vitellius? male. 2. Showing the under side. 3. The supposed female (figured by Mr. Stephens from the English specimen as P. Bucephalus).

Fig. 4. Pamphila Sylvanus, male. 5. The female. 6. Showing the under side.

PLANTS.—Figs. 7 and 8. Aira alpina (smooth Alpine hair grass).

P. Vitellius? is from a specimen taken in North America by Mr. Doubleday, who is decidedly of opinion that the insect captured in England, and described as P. Bucephalus, is the female of this species. P. Sylvanus, male and female, are from strongly-marked British specimens, also furnished by Mr. Doubleday. I have not been able to give a figure of the caterpillar; for, although this species is so generally distributed over the whole island, the caterpillar is at present unrecorded. H. N. H.

SPECIES I.—PAMPHILA BUCEPHALUS. THE GREAT-HEADED SKIPPER.

Plate xl. fig. 1—5.

SYNONYMES.—♂ *Hesperia Vitellus?* Fabricius, Ent. Syst. iii. a. p. 327. Abbot and Smith, Ins. of Georgia? *Papilio Vitellus*, Haworth, Ent. Trans. t. 334. *Pamphila Vitellus*, Stephens. ♀? *Pamphila Bucephalus*, Steph. Ill. Haust. vol. 1. pl. 10. fig. 1, 2, vol. 4, p. 383. Wood, Ind. Ent. t. 3, fig. 82.

It is a curious circumstance that this evidently North American insect should have been captured several times, at distant periods, and in remote parts of this country.

The short Latin character given by Fabricius of his species Vitellus is as follows:—"Alis divaricatis fulvis, anticis macula media marginque, posticis limbo fuscis," and will agree with the supposed male; but his more detailed character, "antennæ annulatæ," and "alæ omnes fulvæ immaculatæ," and his habitat "in Americæ meridionalis insulis," will not exactly accord with it. I have therefore, considering the great similarity of so many of the species, preferred to give the North American species which has been captured in this country, under the name of Bucephalus of Stephens, considering with Mr. Doubleday that the specimens which in this country have been described under the name of Vitellus may perhaps be its males.

The head and thorax of the males are clothed with greenish, fulvous hairs; the front of the antennæ and of the club is fulvous, the hind part brown; the fore wings are tawny above, with slender, black veins. In the centre of the disk is a large, black oval spot, the anterior part of which, as well as the base within, is velvety, the remainder silky; the outer margin is broadly brown, and uninterrupted, although irregularly notched within, two small, connected, transverse, fulvous spots near the tip separating part of the dark border (which thus forms an almost insulated subovał spot, running further into the disk of the wing) from the rest. The hind wings above are darker tawny, with black veins, and a broad, irregularly-notched, dusky border all round the wing, broken near the anal angle by a longitudinal streak of orange, running to the margin. Beneath, the wings are paler tawny, the base of the fore wings black, and the tips slightly brown, preceded by two small, transverse patches of paler buff colour, the upper one being farthest removed from the tip of the wing. The hind wings are marked along the margin with some very slight dusky spots, indicating the dark border of the upper side; and there is a slight dusky spot in the middle of the disk. The under sides of the head and breast are pale buff. The expansion of the wings is a little more than an inch and a quarter. This description will accord with that given by Mr. Stephens of Vitellus, which species is stated by Mr. Haworth, in the Entomological Transactions, (p. 334,) to have been caught in Bedfordshire by the Rev. Dr. Abbott, although he added that he possessed specimens of the same from Georgia. My description is made from a North American specimen in my collection.* With the exception of Mr. Curtis's statement, that "he believes Mr. Hatchett has a pair which he purchased," no other instance is recorded of the capture of Vitellus in this country.

Of Bucephalus, however, or the insect which is regarded by Mr. Doubleday as the female of the supposed Vitellus, two specimens were taken in the neighbourhood of Barnstaple in Devonshire by W. Raddon, Esq., and communicated to Mr. Stephens, who published a figure of this presumed species. In his Catalogue however, (Haust. p. 28,) and in the appendix to his fourth volume, he indicated these as males. His figures, however,

* It is proper to observe that the Vitellus of Abbot and Smith comes nearer to the Fabrician description of Vitellus than that given above. The specimen represented in our figure appears identical with the Augias of the Linnean and British Museum Cabinets. But the Augias is described as a native of Java and India. The Augias of Donovan's Indian Insects must surely be a distinct species.

appear to be those of a female, (which he has in fact informed me is the case,) differing only from the description above given of the male in such characters as are common to the rest of the females in this group of skippers, namely, the want of the discoidal black patch on the fore wings (in the place of which is an oblong dusky spot,) and the more decided and more extended markings on the wings which are only very slightly indicated in the male The large head, the peculiar pointed [illegible] of the marginal spots of the fore wings, and the longitudinal, slender, fulvous streak near the anal angle of the hind wings seem to indicate the sexual identity of Bucephalus with the male described above*, moreover, Mr Stephens "could not avoid surmising that the origin of Bucephalus is questionable and that the specimens were probably imported in one of their earlier states among the timber or other stores, which Mr Raddon acquainted him came direct from the *North American Continent*, to Barnstaple, the section of the genus to which this insect belongs being without any other exception exclusively found in America We have likewise the authority of Mr Doubleday whose acquaintance with North American Lepidoptera is superior to that of any other living entomologist, for considering the two insects to be thus identical It should, however, be added, that Curtis and Stephens (Append vol 4, p 383) mention that Mr Newman had also taken a male of Bucephalus near Godalming This specimen has been figured by Mr Wood in his Index Entomologicus†

SPECIES 2 —PAMPHILA SYLVANUS THE LARGE SKIPPER

Plate xl fig 1—6

SYNONYMES —*Hesperia Sylvanus*, Fabricius, Villars, Gmelin, Ochsenheimer, Curtis *Papilio Sylvanus*, Hubner Lewin Pap pl 46, fig 1—3 Donovan, 8, pl 254 fig 2 Haworth Harris, Aurelian, pl 12 fig 1 *Pamphila Sylvanus*, Fabricius (Gloss), Stephens Duncan, Br Butt 2, pl 2, fig 1 Wood, Ind Ent t 3, fig 80

This, which is the largest of our British Skippers, sometimes measures nearly an inch and a half in the expanse of the wings The upper wings are tawny brown above, with black veins, the costa, a spot on the middle, and an oblique bar beyond the middle, consisting of spots of varied size, marginate behind, and extending nearly to the tip, the two small upper ones being near the margin, whilst three other small spots connected together towards the front margin form with the preceding a very irregular, curved fulvous bar The male has the base of the wings brighter orange than in the female, and an oblique central black patch of hairs The hind wings are dark tawny above, (darker in the female,) with an oblong discoidal, and irregular submarginal row of paler spots On the under side the wings are paler tawny, with a greenish tinge, the anterior at the base and the anal angle in the posterior brighter fulvous, the former with the base internally black The pale spots on the upper side are here represented of a buff colour, but smaller in size The antennae are annulated, the club dark behind, pale in front, the latter has the tip very sharp, and bent into an acute and sudden angle

This common species appears at the end of May, and again at the end of July It frequents the borders of woods, lanes, &c , and occurs in most parts of the country

* The distinction between this male and our figures 1 and 2 must not be overlooked, nor the confusion which exists amongst these Skippers owing to the number of closely allied species It is on this account that I have abstained from speaking more decidedly of the species in question

† Mr Newman has however informed me, since this sheet has been on the press, that he believes his [illegible] Bucephalus to be an Illinois insect, and came with various others he used to receive from Wandborough, Edwards County, Illinois, U S

DESCRIPTION OF PLATE XLI

INSECTS.—Fig. 1. Pamphila Comma, male. 2. The female. 3. Showing the under side. 4. The Caterpillar.

Fig. 5. Pamphila Actæon, male. 6. The female. 7. Another specimen.

Fig. 8. Pamphila Linea, male. 9. The female. 10. Showing the under side. 11. The Caterpillar. 12. The Chrysalis.

PLANTS.—Figs. 13 and 14. Ornithopus perpusillus (Bird's-foot).

The Pamphila Commas are from remarkably well-marked specimens in the British Museum, the caterpillar from Hubner. P. Actæon, 5, 6, are from Hubner's figures; 7 is from a specimen taken by myself at Shenstone, near Lichfield, in 1835, where it was in great abundance, but I did not at the time, being a very inexperienced collector, remark that it differed from P. Linea. The specimens of P. Linea are from Mr. Doubleday's collection, and the caterpillar and chrysalis are from Hubner. H. N. H.

SPECIES 3.—PAMPHILA COMMA. THE PEARL, OR SILVER-SPOTTED SKIPPER.

Plate xli. fig. 1—4.

SYNONYMES.—*Papilio Comma*, Linnæus, Haworth, Lewin, Pap. pl. 45, fig. 1, 2; Donovan 9, pl. 295; Hubner, Pap. 795, fig. 479—481.
Hesperia Comma, Fabricius, Ochsenheimer, Curtis, Boisduval, Zetterstedt.
Pamphila Comma, Fabricius (Gloss.), Stephens, Duncan, Brit. Butt. 2, pl. 2, fig. 2; Wood, Ind. Ent. t. 3, f. 81.
Augiades Comma, Hubner (Verz. bek. Schmett.).
Female Hesperia Sylvanus, Jermyn.

This local species bears considerable resemblance to the preceding, but is distinguished by its darker and more varied appearance, especially on the under side, caused by the pearly white spots, the very different shape of the fore wings in the males, and the different form of the club of the antennæ, and the terminal hook of the club. The fore wings measure about one third of an inch in expanse. On the upper side the wings are of a dark, tawny orange, varied with brown, with the veins black. In the males the basal and central parts are tawny, with an elongated, rather narrowed, incrassated black patch, the middle ridge of which is glossy; in the female the basal and middle part of the wings (except the space between the postcostal and median veins, which is tawny) is dusky; the outer margin in both sexes is broadly dusky, with a very irregular, and much broken and curved series of small spots, which are larger, more distinct, and paler-coloured in the females. The hind wings are dusky with the disk obscure tawny, marked obscurely with about five paler spots in the middle and towards the outer angle. On the under side the hind wings and the tips of the fore wings have a greenish tinge, which is brighter in the females; the hind wings marked with eight or nine squarish, silvery white spots, three towards the base (two of which are often confluent), and the remainder forming a much-curved series parallel with the margin. All these spots are emarginate on the outside. The antennæ are annulated.

The larva is obscure green, marked with reddish, and shining; the head black; the neck with a white collar, and a row of black dots on the back and sides. It feeds on Coronilla varia on the Continent. The chrysalis is elongated and cylindrical.

The perfect insect appears in July and August, frequenting chalky districts near Croydon, and on the chalky downs near Lewes, Sussex. The Devil's Dyke, near Newmarket, and Old Sarum, Wilts, are recorded by Stephens as the localities of this uncommon species. In those places, however, it is very abundant. The Rev. W. T. Bree has also taken it near Dover.

SPECIES 4.—PAMPHILA ACTÆON. THE LULWORTH SKIPPER.

Plate xli. figs. [illegible]

SYNONYMES.—*Papilio Actæon*, Esper, Hubner Pap., Tab. 96, figs. 488, 489, male, 490 female.
Hesperia Actæon, Ochsenheimer, Curtis, Brit. Ent. pl. 442. Godart.
Pamphila Actæon, Stephens, Ill. Haust. vol. 1, p. 383. Wood, Ind. Ent. [illegible] 179. Duncan, Brit. Butt. vol. 2, p. 121 (not figured).
Thymelicus Actæon, Hubner, Verz. bek. Schm.

The expansion of the wings in this species is about an inch in the female, or rather more. The male has the wings on the upper side dusky, with the disk glossed with tawny-orange, the veins black, the fore wings having the usual black oblique dash; occasionally there is a dusky patch at the extremity of the discoidal cell of the fore wings, the tawny colouring beyond this assuming the appearance of a curved series of spots. The under side is more uniformly orange, the disk of the fore wings in the female being more tawny orange; beyond the dark extremity of the discoidal cell is a curved series of six or seven orange spots, divided from each other by the veins of the wings. The under side in this sex has a pearly ochre lustre, a large orange patch on the fore wings extending to the tip of the discoidal cell, where the pale row of spots again appears but more obscurely, and an oblique portion of the inner edge of the hind wings yellowish orange.

In its general character, and the almost uniform colouring of the male, this species approaches *Linea*, but the more maculated appearance of the female approaches nearer to the preceding species.

This extremely local species was discovered in August 1832 by J. C. Dale, Esq., at Lulworth Cove, in Dorsetshire, in considerable numbers, frequenting thistles. It has since been found by the Rev. J. Lockey near the Burning Cliff, in Dorsetshire, in plenty. Mr. Humphreys mentions above that he took it in 1835 at Shenston, near Lichfield, where it was in great plenty.

SPECIES 5.—PAMPHILA LINEA. THE SMALL SKIPPER.

Plate xli. fig. 8—12.

SYNONYMES.—*Hesperia Linea*, Fabricius, Ochsenheimer, Leach, Curtis, Boisduval.
Papilio Linea, Haworth, Donovan, vol. 7, pl. 236, f. 2, male. Harris, Aurelian, pl. 2, f. 1.
Pamphila Linea, Fabricius, Gloss. Stephens. Wood, Ind. Ent. [illegible] f. 78. Duncan, Brit. Butt. [illegible] pl. 1, f. [illegible]
Thymelicus Linea, Hubner (Verz. bek. Schm.).
Papilio Thaumas, Esper, Lewin, Pap. pl. [illegible]
Papilio Comma, Barb.
Papilio Flavus, Muller.

This common little species varies in the expanse of its wings from an inch to an inch and a quarter. The wings above are fulvous, with the veins brown, and a dark margin; the male is distinguished by the ordinary oblique line of black scales on the disk of the fore wings, which is wanting in the female, in which sex the ground colour is not so bright, and the dark margin more suffused within; like the male, however, this sex is destitute of the maculations observable in the preceding species. Beneath, the wings are almost of a uniform colour. The fore wings beneath are paler than above, the base brownish, and the margins pale; the hind wings are ashy-fulvous, with a large fulvous spot at the anal angle. The club of the antennæ in this species is nearly straight and not hooked at the tip.

The Caterpillar is solitary, of a deep green colour, and unspotted, but having a dark line down the back, and two whitish lateral lines margined with black. It feeds on the mountain [illegible] grass and other grasses. The chrysalis is enclosed in a slight cocoon, and is of a green colour.

This species is one of the commonest of the family—flying about low bushes at the outskirts of woods, making its appearance in the beginning and middle of July, and middle of August, and appearing to be distributed all over the country.

SUPPLEMENT

DESCRIPTION OF PLATE XLII

INSECTS — Fig 1 Colias Myrmidone, male 2 The female 3 Showing the under side

,, Fig 4 Hipparchia [Oeneis] Mnestra of Hubner 5 Showing the under side

,, Fig 6 A variety of Hipparchia Janira

,, Fig 7 A splendid variety of Argynnis Lathonia

The Colias Myrmidones are from the figures of Hubner which I have given in this supplemental plate, as it has now some claim to be considered a British species, Mr Stephens having a specimen in his possession which he took near Dover The Hipparchia [Oeneis] Mnestra having been (although without decisive authority) stated to have been taken in England [see ante, p 78] I have thought it interesting to give a figure in this place The variety of Janira [p 70] is from a singular specimen [remarkable for the confluence of the discoidal patches on the fore wings] in the British Museum And the Argynnis Lathonia is a splendid variety, sometimes taken on the Continent, which industrious collectors may hope to meet with in this country H N H

COLIAS MYRMIDONE

Plate xlii fig 1—3

SYNONYMES —*Papilio Myrmidone*, Hubner, Pap fig 432 433 Linst, 1 pl 78, Suppl 21 fig 111, a b Ins (Lepidoptera, Esper) | *Colias Myrmidone*, Godart, Boisduval Icones pl 9, fig 1 2 Ochsenheimer

This species is closely allied to C Edusa (p 15, pl 2, fig 1, 2, 3, 4, and 8), but it is about one-fifth smaller, the wings are rather more rounded of a much brighter orange colour, the posterior, especially, having a decided purplish tinge, the dark border is nearly as in Edusa, but is *never divided at the extremity* of the fore wings *by the slender yellow lines* which are seen in Edusa, on the contrary, it is generally finely powdered with greenish (or, as in Mr Stephens' specimen, with yellow) atoms The under surface of all the wings exhibits nearly the same character as the Edusa The female is rather larger than the male, of a rather duller hue, but decidedly more orange in its tint than the female of Edusa, the dark margin marked with brighter yellow spots, the costal spot pupilled with yellowish—a character which is sometimes also found in the male as well as in some varieties of Edusa

This species, according to Boisduval, inhabits Syria, Hungary, and South Russia, where it flies with Edusa and Chrysotheme, but keeping as distinct from them as Brassicæ, Rapi, and Napi do It has been supposed to be also found in France, but erroneously

This species is introduced on the authority of a specimen in the collection of J F Stephens Esq captured by himself in 1819, between Dover and Brighton, and which he has ever since placed in his cabinet with a ticket "Edusa? var?"

It is proper to add that, having examined the Linnæan Cabinet, I find that the Colias preserved therein, attached to the label of "Electo," (subsequently altered to Electra in the printed work of Linnæus,) is the male of a species closely allied to our Edusa, in which the dark border of the fore wings is not divided by the orange veins, and the silver spot on the under side of the hind wings is very small, with a very minute brown dot attached to it Moreover, specimens of our Edusa are attached to a label also in the hand-writing of Linnæus, marked "Pteridis," and on referring to the works of Linnæus we find no such species, but P Palæno described with the "habitat in Pteride Aquilina" It is to be feared that some confusion has been introduced into the arrangement of these insects *

* I also found Chrysophanus Chryseis attached to the Linnæan label of Hippothoe

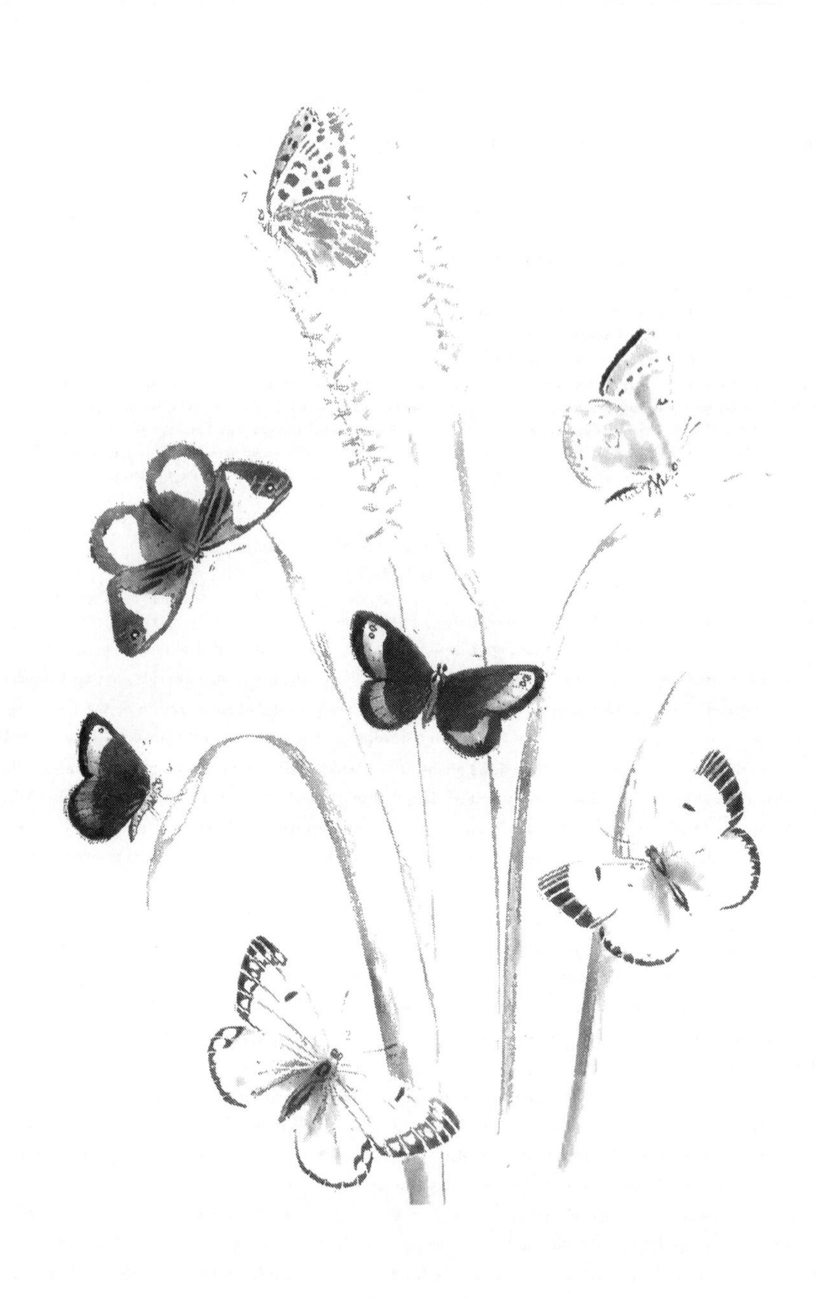

DIRECTIONS FOR COLLECTING AND REARING THE CATERPILLARS, AND PRESERVING THE PERFECT INSECT.

HAVING in the preceding pages given a portrait and description of every species of Butterfly indigenous to Britain, or reputed British, and also the larva and pupa of each, as far as they are known, it only remains, in conformity with a promise in my preface, to add a few suggestions as to the best mode of collecting and also of rearing caterpillars from the egg, and the most approved manner of setting out and preserving the perfect insect.

To become a fortunate collector, requires not only much industry in the pursuit, but also a keen observation and study of natural phenomena in general. For instance, many persons, with all the necessary enthusiasm and industry, and perhaps a great sacrifice of time, take numerous collecting excursions with scarcely any success; whilst others, with less eagerness and much less expenditure of time, seldom return without numerous captures. The cause lies in a proper selection of season and weather for the objects in pursuit.

It is almost useless to attempt collecting winged insects during a cold east or north-east wind, and places at other times abounding in insect-life will then be still, and to all appearance deserted. A warm and genial day is therefore, above all things, necessary, and to secure this desideratum it is necessary to become as far as possible *weather-wise*. Mr. Ingpen, in his excellent little work upon collecting insects, mentions many circumstances which different writers have considered infallible *signs* of fair weather; but he considers most of them doubtful—such as the opening of the pimpernel, the early flight of the cabbage-white butterfly, &c. &c.; whilst he considers the *high* flight of swallows an almost certain forerunner of a fine day. Sir H. Davy, in his delightful "Days of Fly-fishing," has philosophically accounted for this and many other natural phenomena which have become popular omens. "Swallows (he says) follow the flies and gnats, and flies and gnats usually delight in warm strata of air; and as warm air is lighter and usually moister than cold air, when the warm strata of air are high, there is less chance of moisture being thrown down from them by the mixture with cold air; but when the warm and moist air is near the surface of the earth, it is almost certain, as the cold air flows down into it, a deposition of water will take place." As these instructions are intended for beginners, and not for the accomplished entomologist, I may usefully add a few more such remarks from the same source. "It is always unlucky (for anglers in spring) to see a single magpie—but *two* are a good omen; and the reason is, that in cold and stormy weather one magpie always remains sitting upon the eggs or young ones to keep them warm; when the two go out together, the weather is warm and settled." Another popular sign of fine weather, is, when the red clouds of the setting sun take a tint of purple; upon which the same author remarks that "the air when dry refracts more red, or heat making rays; and as dry air is not perfectly transparent, they are again reflected on the horizon I have generally observed a coppery or yellow sunset to foretell rain, but as an indication of wet weather approaching, nothing is more certain than a halo round the moon, which is produced by precipitated water and the larger the circle, the nearer the clouds, and consequently the more ready to fall" I must not omit, in conclusion his beautifully simple verification of the rustic couplet,

"A rainbow in the morning, is the shepherd's warning
A rainbow at night is the shepherd's delight"

'A rainbow can only occur when the clouds containing or depositing the rain are opposite the sun, and in the evening therefore the rainbow is in the east, and in the morning in the west, and as our heavy rains in this climate are usually brought by westerly winds, a rainbow in the west (occurring only in the morning) indicates that the bad weather is on the road, by the wind to us, whilst a rainbow in the east (occurring only in the evening) proves that the rain in these clouds is passing from us'

These remarks of the philosophic fly-fisher, beside the general information they convey, may teach the young entomologist how to select his weather with a good chance of a fine day, and also that popular omens are not to be rejected at once by superficial and pert reasoning, but that they are generally founded on truths, however deeply concealed by an accumulation of fancy or superstition Having stated that a favourable day is indispensable to a successful search for insects, more particularly as I am now referring principally to butterflies, as the only class of insects treated of in this volume, it is next necessary to suggest the best seasons for search Long then before any specimens are to be taken in the winged state, the collector may, as early as the end of January, dig for the chrysalides of such species as enter the earth to undergo their transformation to the pupa state, and these he will be most successful in finding near the roots of such plants as the caterpillars feed upon Other species he will find upon walls or paling, near the food of the larvæ, and others still attached to the withered stems of the plants of the previous summer He may also search for the eggs of many species, the most likely places to find which, and the various modes of depositing them, he will find described in the body of the work Caterpillars may be collected as early as the beginning of April, and the best time to find them is early in the morning and late in the evening, or even night, as many species remain concealed during the greater part of the day, and some feed only at night, consequently a search for them by day would be fruitless, though some might occasionally be found by pulling up the plants and carefully examining the roots, about which they sometimes lie concealed Caterpillars may be sought all through the summer, and as some butterflies are what is called double-brooded, their larvæ are to be found as late as September It will be useless here to repeat at what particular season each species is found, as that will be found fully described in the preceding pages I therefore merely remark in addition, that some that can scarcely ever be seen, as they feed at the top of high trees, may be taken by shaking or beating the tree—such, for instance as the beautiful larva of the purple Emperor, which feeds upon the highest branches of the oak, which is indeed a fertile theatre for the occupation of the entomologist, each tree affording shelter and food to various tribes of insects, too numerous to specify A white cloth or sheet should be spread upon the ground before beating or shaking trees [These observations are as applicable to moths as to butterflies]

Whenever the collector is a draughtsman, a careful and exceedingly accurate drawing of several individuals of every species of caterpillar should be taken, and each caterpillar kept separate, and distinguished by a *number*, corresponding with a *number* attached to the drawing, and by this system not only every butterfly will be assigned to its proper caterpillar (which has not always been the case), but even the male and female caterpillars may perhaps be distinguished by unvarying markings, as distinct, no doubt, in many instances, as those of the perfect insects themselves; a fact which it would be highly interesting to prove satisfactorily.

The caterpillars, when taken, should be touched with care, as they will not bear rough handling. A large box should be prepared for them with a gauze lid and should contain several divisions, each distinguished by a *number*; each division should also have a little earth mixed with rotten wood at the bottom, which may be prevented from getting too dry and dusty by keeping a layer of damp moss upon it. In the corner of each division should be placed also a phial of water, in which a branch of the plant which the insect feeds upon will be kept fresh; it should, however, be renewed every day, or even twice a day if possible, care being taken not to disturb the caterpillars at the time they are casting their skin, which occurs several times before they attain their full growth, varying in different species. It will be understood that the earth at the bottom of the divisions is for the use of such caterpillars as undergo their change in the ground.

To rear caterpillars from the egg is much more difficult, but the most certain method is to place the eggs securely upon a branch of the proper food of the species, in the open air, and to prevent escape, inclosing the branch in a gauze or muslin bag or frame. It will be found necessary, however, to remove them to other branches as often as the leaves are destroyed, or become unfit food. Caterpillars, when taken nearly full-grown, may also be treated in this way with great success, but great care must be taken in removing the chrysalides to a box covered with gauze as soon as they are formed, and they must in all cases be examined frequently, as if the perfect insect remains long in the box without being secured, the wings will become injured by its endeavours to escape; and one great advantage of rearing them from the caterpillar state is, that more perfect specimens are secured than could possibly be obtained by capturing them in the winged state, as even the exercise of flying destroys the downy bloom which they exhibit on first emerging from the chrysalis.

To capture the winged insect flying, or settled upon a flower, or on the ground, gauze nets are used of two or three sorts, which will be found described in Mr. Ingpen's little work, or Mr. Westwood's Entomologist's Text-Book; for instance, to capture the high-flying purple emperor a net is sometimes used fixed to a rod or pole twenty feet long; but Mr. Ingpen mentions that he is sometimes, in common with other strong flyers, brought to the ground by throwing up a piece of stone or tile, in his course, which he follows in its fall, and sometimes alights upon it, when he is easily taken.

When captured and killed, care being taken not to rub off the down from the wings, a pin must be passed through the thorax, and the wings kept expanded by thin braces of card until the insect is perfectly dry. This requires several days, varying according to the weather, after which it is ready to be placed in the cabinet. When the season is past, both for taking the insects in the larva or imago state, the leisure hours of late autumn may still be occupied in search of chrysalides. These may be sought as the garden flower-beds are dug over, upon the plants on which they have fed, and on walls and palings; but in the latter situations it frequently happens that they are diseased individuals, which, pierced by the ichneumon fly, have wandered from their food and in their *malaise* sought a shady and solitary retreat, instinctively perhaps, endeavouring to escape their enemy, who generally pierces them in the bright sunshine. Chrysalides taken in such situations will frequently, when they burst, instead of the expected butterfly, discharge a hundred small silken cocoons, each containing a chrysalis, from which eventually issues a small fly, which, in its turn, seeks some unhappy caterpillar, and, by means of its sharp ovipositor places a number of eggs in its body, which, quickly hatched by the warmth, feed upon its vitals till it is destroyed The ichneumon of the small silken cocoons, mentioned above, seems to confine its ravages to the caterpillars of the cabbage-white butterfly, but each species has its peculiar foe of this description some large and some small, the former depositing only one or two eggs in the body of each caterpillar, the latter from ten or twelve to near a hundred I have seen a caterpillar of the lacquey moth wince under the repeated punctures of its ichneumon foe, till it has at last fallen from the branch upon which it was feeding, it, however, soon resumes its food, doubtless with redoubled rapacity to satisfy the insatiate legion within, till, overcome by exhaustion, it crawls away to fix itself in some solitary place, where the chrysalis is found

After the season of collecting is entirely over, or when bad weather confines the student to the house, he may occupy his leisure time in arranging his collection, and I would strenuously advise him to do so, not merely as a pretty display of beautiful objects, but with due regard to nomenclature and system Doubtless the most deeply interesting portion of natural history is the observation of the habits, physiology, structure, and properties of organised creatures (by far the greater number of which belong to the entomological division), but then proper and convenient arrangement, according to the most recent terms and system of science, is absolutely necessary for the successful progress and application of all knowledge, and even those who are confining themselves to the arrangement of the mere nomenclature of the catalogue, are doing good service to the advancement of the science The elaborate and searching observations of Reaumur and Bonnet would have been much more valuable had they been conducted with such a view to system and arrangement, whilst as it is, (as mentioned in Kirby and Spence's Introduction) some of the insects of which they have recorded the most interesting circumstances, cannot, from their neglect of system, be at this day ascertained No one, for instance, knew Reaumur's *Abeille tapissière* until Latreille, happily combining system with attention to the economy of insects, proved it to be a new species *Megachile Papaveris* Even with the assistance of carefully-coloured portraits, it is almost impossible so to describe insects as to render them recognisable with certainty, unless the accepted terms and system of science be also employed and Mr Westwood, in his Entomologist's Text-Book alludes to the fact that many rare insects, of which engraved portraits have been given by the early entomologists, have from this cause been thought to be new species

Kirby and Spence affirm that a well-arranged system, with proper terms and names is as necessary to the understanding of a science as is a dictionary to the understanding of a foreign language "The labours of a Michaelis or a Laplace might be sealed books to us without dictionaries of the French and German languages, and in fact a good system of insects, containing all the known species, arranged in appropriate genera families, orders, and classes, is in reality a dictionary, enabling us (without the incalculable loss of time which would otherwise occur) to ascertain the name of any given insect, and thus to learn all that has been recorded of its properties and history, as readily as we determine the meaning of a word in a lexicon

As far as regards the systematic position of any insect connected with the order treated of in the present volume, the student will have, comparatively, little trouble The veriest tyro will at once know that any *butterfly* or *moth* must belong to the order *Lepidoptera* He will perceive that as a *butterfly* it must belong to the section *Diurna*, a few striking characteristics will show him to which family he must refer it, and it then only remains to ascertain its genus, and supposing it to belong to the genus *Vanessa*, it will not be at all difficult to ascertain the specific name, as the different species of this genus are at once obviously distinguishable by their various markings alone without reference to their minute structural characters, which should, however, always be attended to by the student. If, after pursuing this course he finds he has an insect evidently of the genus Vanessa, or any other, but that it records with none of the recorded species he may hope to have been the discoverer of a new species particularly if he reared it from the caterpillar, and, (having procured an accurate drawing of it) if he finds it differ from that of the species in question, he *must then from its* characters the differences of structure or marking seek to give it such a specific name as will be acknowledged by science and will serve to distinguish it from the rest of the genus, and also describe it by such a *character* as he will find at the head of every species in this work. It is true that in butterflies, which besides their conspicuous colouring, fly at high noon, the collector can hope to make few discoveries of this description; nevertheless several such have occurred even within the last three or four years and when we consider that the beautiful Lycæna dispar was only discovered about the year 1822, there are doubtless still some novelties in store for the industrious collector, even among our butterflies; but among our moths, of which I am preparing a series of similar illustrations to those of the butterflies, very numerous discoveries may be expected, for, flying at the dead of night, or at the early dawn, many must at present have escaped the search of entomologists, particularly among those which appear in the winter when the collector is seldom abroad.

In conclusion, I have only to say that if the reader has any doubts respecting the utility and importance of entomology as a science, let him not only inquire what eminent men have devoted a large portion of their lives to its pursuit, but let him at once read Kirby and Spence's Introduction to their beautiful work upon the subject one of the most interesting and convincing pieces of writing in the language—and let him there learn what light has been thrown by the science upon the labours of the silkworm, whose product furnishes labour and subsistence to millions, upon the ravages of the turnip-fly, or the instinctive mechanism of the bee or the ant, and upon those links which it furnishes to the great chain of organisation and intelligence, from infinite perfection, to the brink of dreary nothing.—H. N. H.

ADDENDA AND CORRIGENDA

P 8, Note *—Mr Stephens has suggested to me that Mr Curtis's figure represents the variety of Podalirius regarded as distinct by Boisduval under the name of Duponchelii, and certainly, so far as the description of Boisduval goes it records therewith, but that author has not noticed the much greater extent of the black markings which led me to infer, from the generally adopted doctrine, that this darkness in colour was attributable to the more northern locality of Mr Read's individual, which thus appeared to be confirmed as a native specimen Mr Curtis's figure, in fact, represents a specimen with the ground colour of the wings whiter than ordinary, the costa and marginal lunules of the hind wings fulvous, and the anal ocellus very distinct in its markings If this be the true character of the South European and African varieties of Podalirius we must deny the indigenousness of Mr Read's specimen

P 18, Note *—Omit the latter paragraph of this note

P 19, line 6—*for* "omitted" *read* "emitted"

P 28, last line—*for* "Sinapis" *read* "Candida"

P 32, line 19—*for* "Renner" *read* "Rennie"

P 32 line 23—The specimens in which the apical patch is entirely wanting have been considered as a distinct species by Borkhausen, under the name of P Erysimi

P 36—The British Fritillaries, in respect to the arrangement of the wing-veins form two groups 1st The Argynnæ including Lathonia, in which the postcostal vein of the fore wings emits *two* branches before joining the transverse vein, and a third branch at the junction of the transverse and postcostal veins, this third branch emitting *two* branchlets, whereas in 2nd the Melitæa, the postcostal vein only emits *one* branch before jointing the transverse vein, and a second at the junction of the transverse and postcostal veins, this second branch emitting three branchlets extending to the costa Without an entire revision of the whole group of Fritillaries, it is impossible to determine whether this character is of higher value than those which I have suggested, and which separate Lathonia from the other Argynnes, with which it is united by means of the new character described above, and which, I need scarcely add is now for the first time introduced

P 36 line 26—*for* "Lattuna" *read* "Calluna"

P 43 line 4—*for* "10" *read* "8"

P 47 last line but one—*for* "probably" *read* "properly"

P 58, line 18—*for* "Levanca" *read* "Levana"

P 71, lines 22 and 25—*read* Cœnonympha

P 73, line 17—*for* "Wales" *read* "Wales"

P 100, line 5, *for* "Cymon" *read* "Acis"

P 107, line 21 *for* "brown" *read* "broom"

The following additional localities are given by Mr Henry Doubleday in "The Entomologist" for August 1841 (Vol I p 156) —

Aporia Crataegi Plentiful in Monkswood Hunts 3rd of June

Thecla Pruni, first appearing on 18th of June in Monkswood

Polyommatus Arion A single male near Wigsworth Northamptonshire, in June It is a singular variety, and [illegible] larger than P Argus Can this be one of the nearly allied Continental species?

Melitæa Artemis In Monkswood, Holme fen, and in profusion near Aldwinkle in Northamptonshire

Chlorophanus dispar Caterpillars very plentiful in Holme fen, on the Water Dock (Rumex Hydrolapathum)

In addition to the species described in the preceding pages, the following have also been incautiously introduced into the lists of British species, but upon such slight authority that it has not been deemed necessary to figure them in this work

PARNASSIUS MNEMOSYNE, Latreille (Papilio Mn Linnæus) It is smaller than Apollo, from which it is at once distinguished by wanting the ocelli The veins are slender and blackish, the fore wings with two black spots in the discoidal cell It inhabits the Alps, Pyrenees, Switzerland, Sicily, Sweden, Hungary and Russia It may therefore possibly still be found in England It was introduced by Turton and Jermyn, and is figured in Wood's Ind Entomol pl 53 f 4, amongst the doubtful British species

PIERIS FERONIA (Pontia F Stephens, Ill Haust, 1, 140 Ernst, Papillons d Europe, vi p 209) Wings above white, the anterior with a single row of triangular brown spots touching the hinder margin, and terminating in a point on each nervure internally, beneath immaculate, the anterior white, with a yellowish tint on the outer angle, the posterior entirely of the latter colour irrorated with dusky ' Ernst, by whom alone this species appears to have been noticed and figured amongst Continental authors, says of it that it was "prise en Angleterre' No other authority exists for its being an indigenous species, and Mr Stephens suggests that it may be a native of New England, in America

MELITÆA THAROS (Papilio Dan Fest Tharos Drury, App v 2, Cramer, pl 169, fig E F, Argynnis Tharossa, Enc Meth 9 289, Melitæa Tharos, Westw in Drury, 2nd Edit 1, p 39) The wings are black brown, with many orange marks, some of which form an irregular bar beyond the middle of the fore wings, the tips and margins being dark There is also a row of black round spots in orange spaces beyond the middle of the hind wings This is a common North American insect, but Cramer, in figuring it, stated it was "reçu d'Angleterre," whence it has been inferred that it was regarded by him as an English species

LIMENITIS POPULI Fabr (Papilio P Linn Stewart, Wood, Ind Ent t 53, fig 10) The wings above brown, fasciated and spotted with white beneath, luteous, fasciated with white, and ornamented with blue spots This fine species, which is nearly three inches in expanse, appears to have been introduced in the English lists in consequence of Linnæus having erroneously referred to Ray's description of Camilla, amongst his synonyms of Populi

LIMENITIS SIBILLA, (Papilio S Linn Stewart, Wood Ind Ent t 53, fig 11, L Camilla Fabricius), is closely allied to L Camilla, with which it has been confounded by Fabricius Stewart, who followed the nomenclature of the latter author accordingly gave Sibilla as a native species The true Sibilla is nearly two inches and a half in expanse, the wings above, dark brown, with a white fascia, without any red spot at the anal angle, beneath, orange tawny, spotted as above No authentic instance is recorded of its capture in this country

POLYOMMATUS TITUS (Hesperia Titus, Fabricius, Ent Syst 3, a p 297, Turton Pol Titus, Jermyn, Stephens) Habit of Argus and Artaxerxes All the wings above brown, unspotted Beneath also brown, the anterior with a hinder row of white and black lines, the posterior with a short central line, and a row of black spots ocellated with white Near the margin are a row of red spots, each marked with a black dot "Habitat in Anglia Dom Drury Jones fig pict 6, t 41, t 2' It appears that Fabricius derived his knowledge of this species from the same source, whence he also described Artaxerxes, namely, Jones's Collection of Drawings Beyond this we have no information respecting the species

ALPHABETICAL INDEX

OBS.—*The ordinary Synonyms of the Genera are inclosed in brackets, and those of the Species are printed in Italic.*

LONDON
BRADBURY AND EVANS, PRINTERS,
WHITEFRIARS.

Lightning Source UK Ltd.
Milton Keynes UK
UKHW020019100223
416721UK00002B/354